A BRIEF HISTORY OF
ALBUM COVERS

Publisher and Creative Director: Nick Wells
Project Editor and Picture Research: Sara Robson
Art Director: Mike Spender
Layout Design: Theresa Maynard
Digital Design and Production: Chris Herbert

Special thanks to: Nick Bennett, Frances Bodiam, Tim Bodiam, Chelsea Edwards, John Esplen, Victoria Lyle,
Fiana Mulberger, Steven Robson, Julia Rolf, Chris Stylianou and Claire Walker

First published 2008 by
FLAME TREE PUBLISHING
Crabtree Hall, Crabtree Lane
Fulham, London SW6 6TY
United Kingdom

www.flametreepublishing.com

Music information site: www.musicfirebox.com

10 12 11 09

3 5 7 9 10 8 6 4 2

Flame Tree is part of The Foundry Creative Media Company Ltd

All images courtesy of Private Collection/Foundry Arts

The CIP record for this book is available from the British Library.

ISBN: 978-1-84786-211-2

Printed in Singapore

A BRIEF HISTORY OF
ALBUM COVERS

JASON DRAPER
FOREWORD BY PAUL DU NOYER

FLAME TREE
PUBLISHING

CONTENTS

6

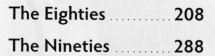

HOW TO USE THIS BOOK

The reader is encouraged to use this book in a variety of ways, each of which caters for a range of interests, knowledge and uses.

- The book is organized by decade into six chapters: **Fifties**, **Sixties**, **Seventies**, **Eighties**, **Nineties** and **Noughties**.
- Each chapter is then further organized chronologically according to the earliest release date of the album, either in the UK, US or occasionally Australia.
- Each chapter contains a selection of some of the decade's most iconic, unusual or representative album covers.
- Each entry is designed to provide information about an album cover and the context in which it was created.
- **Record Labels:** this information relates to the record labels responsible for the album's initial release only. If first released on different labels in the UK and US, both are given.
- **Release Dates:** where an album's initial release date differs between the UK and US, both dates are given.
- **Songwriters:** the names of the songwriters credited on each album are located at the bottom of the page, and are separated into principal and secondary songwriters where necessary.

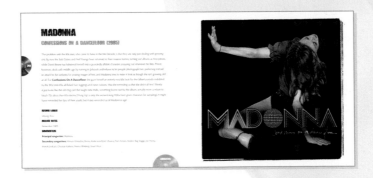

Name of the artist or band

BOB DYLAN

SELF PORTRAIT (1970)

Album title

Earliest release date

The album in which Bob Dylan tried to lose his fans. In 1969 he had become the first-ever bootlegged artist with **Great White Wonder**, a collection of unreleased studio outtakes. Lying low at the end of the 60s, Dylan thought if they wanted outtakes, he would give them outtakes, and hopefully they would leave him alone. However, loading up a double album with below-par live material, knock off covers and a few lazy songs of his own backfired, and the clamour for the 'real' Bob Dylan to return grew stronger after he presented his fans with stuff they did not, in fact, want. **Self Portrait** flopped, but it was one big joke to Dylan anyway, who wasn't *really* saying this was the real him. 'There was no title for that album,' he said in 1984. 'I knew somebody who had some paints and a square canvas, and I did the cover up in about five minutes. And I said, "Well, I'm gonna call this album **Self Portrait**'" Hardly looking a thing like Dylan, in this case you *can* judge an album by its cover: knock-off, rushed, and an attempt by the artist to get people off his back.

Information about the cover and the context in which it was created

RECORD LABELS

CBS, Columbia

Record label(s) responsible for the initial release (UK, US)

RELEASE DATES

July 1970 (US: June 1970)

Initial release date(s)

SONGWRITERS

Principal songwriter: Bob Dylan

Secondary songwriters: Gilbert Becaud/M. Curtis/Pierre Delanoe, Felice Bryant/Boudleaux Bryant, Alfrid Frank Beddoe, Paul Clayton/Larry Ehrlich/David Lazar/Tom Six, Lorenz Hart/Richard Rodgers, Gordon Lightfoot, John Lomax/Alan Lomax/Frank Warner, Cecil A. Null, Paul Simon

Songwriters credited on the album

SEVENTIES

9

FOREWORD

'Brown paper bags for **Sgt. Pepper…**,' begged Brian Epstein. The Beatles' manager was horrified by his boys' brainwave – an extravagant gatefold sleeve by the Pop Artist Peter Blake. The idea looked ruinously expensive, it was a nightmare to organize, and the array of famous living people – not to mention infamous dead ones – presented endless potential for legal grief and moral outrage. But the Beatles got their way, and what a good thing too.

There had already been well-designed record sleeves, of course. But after **Sgt. Pepper…** the business of putting protective packaging around recorded music would never be the same. Suddenly, the functional album sleeve became nothing less than an art form. It is no exaggeration to say that many an album's wrapping has proved more enduringly interesting than the music itself.

Illustrated sleeves had been standard since the arrival of LPs (long-playing vinyl discs) in the 50s. Colourful 12-inch squares replaced the more prosaic dust covers that enclosed the LP's predecessor, the 78 rpm disc. (In fact, they revived an almost dormant art form, namely the covers of sheet music sold in shops and music halls since Victorian times.) The idea of an 'album' of songs, coherently themed, was made popular by Frank Sinatra. In 1958 **Come Fly With Me** depicted Frank enticing us aboard a global trip on wings of song; its upbeat sleeve image (featuring the TWA Jetstream Super Constellation) was sponsored by an airline company. In the same decade, jazz labels were especially quick to seize upon the album sleeve's possibilities: the cool modernism of the music was advertised impeccably by the crisp design surrounding it.

Despite a few notable high points, most early album covers have only a retro charm. The work of anonymous toilers in art departments, they were normally formulaic, adorned by serviceable mug shots of the performers. It is sometimes more fun exploring the paper inner sleeves, which were festooned with breathless trailers for other artists, or bore stern advice upon correct procedures ('must be played using a sapphire or diamond stylus with a tip radius of between .0005 and .0007 inches'). And sleevenotes were a world of their own, ranging from hack superlatives to Bob Dylan's stream of hipster consciousness on **Bringing It All Back Home**.

Encouraged by the Beatles' evident attention to sleeve quality, the album cover entered a golden age around 1966. No longer was it merely pretty packaging. Suddenly these covers were cultural statements, on which fantasies were projected and unorthodox cigarettes were rolled. By 1966 even Andy Warhol – who had once drawn covers for the Blue Note jazz label – was joining in, bestowing a magisterial banana on the debut by his protégés the Velvet Underground. For the next few decades, album covers ruled as emblems of generational identity. Even their junior cousin, the seven-inch single sleeve, enjoyed a period of intense creativity.

The great challenge came with the advent of compact discs in 1983. Only five inches in diameter, the CD effectively shrank the canvas for sleeve art to less than half its former size. The music industry began transferring its back catalogue to the new medium; immortal designs from an earlier age had to adapt to this unforeseen downsizing as best they might. At least CD packaging was still square, unlike the LP's earlier and even smaller rival the cassette tape. And, happily, there was no mass shift to other new carriers being touted, like the MiniDisc and Digital Compact Cassette.

In the event, designers simply made their peace with new technology and, as we now know, the flow of imaginative imagery never stopped. It is impossible, for example, to think of Nirvana's **Nevermind** without visualizing that underwater infant and the sodden dollar bill. In fact the CD age has abounded in cover ideas of great potency – often as part of the artist's wider aesthetic programme, like the urban cartoon cool of Gorillaz, the confrontational starkness of Hard-Fi or the colour-coded self-branding of the White Stripes.

The danger is that, having shrunk, the canvas may now disappear entirely. Our musical habits are migrating online; the art of the album cover is threatened by the potential extinction of the CD. There is an irony at work here. The irresistible rise of video, marketing and media means that pop is more visually led than ever. But can the physical cover survive? In the digital age you can download albums with artwork intact, yet the impact on the screen of an iPod is meagre indeed. Peer a little more closely and they still look glorious, those covers, even in miniature. But shall we ever see their like again?

So let us celebrate the accidental art form of the album cover. Never conceived as more than a commercial teaser, it somehow became the most exciting visual medium of its time. When the sound met the look, it was a marriage made in heaven. The images might be grungy, glamorous, erotic, funny, dumb, pretentious, strange, or anything else. Collectively, they made for a superb art gallery in some humble corner of every music-lover's home. Perhaps they will ultimately live on real gallery walls, reverently framed. Perhaps future generations will say: 'Apparently they came with music, as well.'

Paul Du Noyer, 2008

INTRODUCTION

In a world where digital music threatens to obliterate the art of album packaging entirely, it is even more important that we remind ourselves of what a singular type of creativity we may be pushing to the sidelines. As downloading music becomes more and more popular, and record labels begin to increasingly rely on it for revenue, producing something physical, with packaging, may soon seem hardly worth it from a fiscal point of view. Sure, you can sometimes receive exclusive 'sleeves' or digital booklets with your downloads, but clicking a few buttons and staring at images passing by on your computer screen is just as unsatisfactory when compared to holding a CD booklet, as holding a 12-centimetre-by-12-centimetre CD booklet is to taking in the sensory overload of a 12-inch-by-12-inch vinyl LP.

THE CARS
CANDY-O

Vargas

As time has told, however, vinyl manufacturing has seen a revival in recent times, and with it a reassessment of the limitless possibilities of album art on a larger canvas. So if CDs couldn't kill it off, then perhaps the digital download will fail as well.

More than just being something nice to look at, album art has a number of uses. Some artists, you will notice, simply use it as a marketing tool, picturing the sleeve up on the racks with all the other new releases and thinking of ways in which their album can stand out from the rest – almost in the way that magazines visually jostle for space in the newsagents. Examples include The Cars' **Candy-O** (see page 195) and Carly Simon's **Playing Possum** (see page 143), both of which rely on sexual imagery to catch the buyer's eye, almost as if selling the album on the back of the music is an afterthought.

Similarly, the likes of Frank Zappa, who appears in this book twice, almost entirely removes his artwork from mirroring the music inside, treating album sleeves as separate entities, featuring an amusing image that would work just as well on its own as it would connected to a piece of music (see **Weasels Ripped My Flesh**, page 91). Delving even further, major artists such as Pink Floyd and Led Zeppelin, very often working with top-quality album art designers (most notably the design company

Hypgnosis, though this book also features works from individual artists such as Robert Crumb, Mati Klarwein, Neon Park, Andy Warhol, Pedro Bell and H.R. Giger), treat their sleeves as a blank canvas on which they can create a genuine piece of high-class art that should be revered as much as the music, but that does not necessarily work in tandem with it.

On the other hand, many bands choose to intertwine their music with their visuals so intensely that it is hard to decide where one begins and the other ends. Radiohead's **OK Computer** artwork (see page 331) sees the band extend the musical and lyrical themes of their album across a whole booklet and sleeve, making sure that the package comes as a meaningful, inseparable whole, as opposed to a collection of songs being sold on the strength of an image. Some even go so far as to question what exactly an album sleeve needs to be, with the Beatles stuffing **Sgt. Pepper's Lonely Hearts Club Band** (see page 49) full of nostalgic paraphernalia such as badges and fake moustaches, and System Of A Down stripping down in the opposite direction, imploring their fans to **Steal This Album!** and offering them

nothing more than a CD-R full of music, with handwritten tracklisting and title (see page 355).

Simple branding crops up time and time again, as you will see with the White Stripes' instantly recognizable red-white-and-black colour schemes, the Rolling Stones' tongue-and-lips logo and even Madonna, whose album sleeves often rely on different blues, making certain shades of pastel blue almost synonymous with her image today (see her sleeve for **True Blue**, page 265, the first of her sleeves to do so). Going hand-in-hand with this is the very idea of creating an image, as bands seek to prove themselves to be fun-loving popsters, the height of stylized cool (Gorillaz' Jamie Hewlett designs proving you really can construct a look from nothing) or the ultimate in bohemian cool (Bob Dylan and Tom Waits' **Highway 61 Revisited** [1965] and **Small Change**, see page 157, respectively). Some bands' entire image has endured thanks to the popularity of just one chance album shot, as evidenced in the iconoclastic sleeves for Elvis Presley's, self-titled debut (see page 21), the Clash's **London Calling** (see page 207) or Bruce Springsteen's **Born To Run** (see page 147).

Equally popular is the idea of image destruction, especially in the early 70s, it seems, where established artists such as Bob Dylan (**Self Portrait**, see page 89) and Pink Floyd (**Atom Heart Mother**, see page 95) kicked against their perceived image by providing something that was the very antithesis of what fans expected (a cow in a field? From a space-rock band? You must think we're stupid …). Of course, David Bowie took this to its logical extension when he destroyed one image only to create a brand new one for himself with each succeeding album.

It is interesting to look at how designers have approached album art throughout the decades. Most 50s jazz sleeves tried to portray some sort of natural cool in the simplest way, while from the mid-60s onwards, especially following the Beatles' **Sgt. Pepper...** in 1967, album art really came into its own as a design phenomenon. The 80s saw an artist's image almost rival their actual music for importance, and if you could prove your conformity to the fashions with a *zeitgeist*-capturing look, so much the better (Cyndi Lauper may have called her album **She's So Unusual**, see page 253, but it essentially establishes the stereotyped fashion for young women across the entire decade). The 90s bucked this trend, as artists fought to prove they had the weirdest, most unexpected sleeve, highlighting their individuality. It worked for a while (Nirvana's **Nevermind**, see page 295, and Underworld's **dubnobasswithmyheadman**, *see* page 313), until even that became a cliché, with far too many collections of random images proving too difficult to differentiate on the shelves.

Of course, some people just stick to good old rock'n'roll 'fuck you's to get their sleeve across. Blind Faith knew what controversy they would cause when they put a pubescent girl on the front of **Blind Faith** (1969), pointing a phallic symbol down towards her crotch, while the Sex Pistols were happy enough to swear their way into public consciousness, with the use of the word 'Bollocks' in the title (see page 177), leading to two stockists of the album being jailed. In hip hop, a genre the media loves to focus on when it comes to 'beefs', you have Ol' Dirty Bastard proudly displaying his welfare

card as his album artwork (he was still collecting welfare cheques two years after **Return To The 36 Chambers**, see page 323, was released, despite the album then being in the Top 10 of the American charts), and Tupac holding up his 'Westside' hand symbol on **All Eyez On Me** (1996), ensuring that every East-Coast rapper in the US knew that he was at war with them.

What is interesting to note is the freedom given to the more established acts at the time. If a brand new band had provided their label with a dog's dinner sleeve such as **Self Portrait**, it would be rejected out of hand. With Bob Dylan, however, it was accepted as an artist's statement. When the Beatles wanted to put out a completely blank sleeve, the only compromise made was that it at least featured their name somewhere (three years later Sly & the Family Stone would not even have to do that for **There's A Riot Goin' On**, see page 103).

In this book we are not necessarily bringing you the most iconic sleeves as you would find in your high-street magazines' 'Top 100 Best Album Covers Ever', though we are featuring a great deal of them. Similarly, we are not necessarily bringing you the sleeves of the 'Top 100 Albums You Must Hear Before You Die', and sometimes we are not even bringing you the album sleeve you may associate most with a given artist. Instead, we have sometimes favoured the leftfield choice, because of the way it fits in (or doesn't, even) with a particular artist's body of work, or perhaps because it provides the moment where an artist's image began to crystallize, proving itself the launching pad for a great many more sleeves to come. Sometimes it is the sleeve that, while not generally remembered, provides a stepping-stone in the general approach to creating album artwork. Some, of course, are just eye-catching images, highly stylized paintings or photos that you just wouldn't get in any other artistic medium. Almost all say something about the band or society at the time, even if they define its subject by what it is not, as opposed to what it is.

We haven't even begun to look at the single 7-inch/EP picture sleeve or the picture disc, all artworks in their own right with their own rules of creation. Perhaps those are for another book entirely. For now, however, from AC/DC to Frank Zappa, high art to low sales, and happy accidents to months and months of painstaking design, this is *A Brief History Of Album Covers*.

Jason Draper, 2008

ALBUM COVERS

RIFF
TORY

FIFTIES

ELVIS PRESLEY

ELVIS PRESLEY (1956)

So iconic that the Clash paid homage to this sleeve for their **London Calling** album (see page 207) – as did Tom Waits for his slightly less well-known **Rain Dogs** sleeve (1985) – **Elvis Presley** came out in 1956, as much a year zero in music as punk's 1977. It showed the record-buying public just what, exactly, rock'n'roll was, with the black-and-white shot of Elvis embodying unbridled teen energy. Having him off-centre and juxtaposed with seated drummer DJ Fontana is a masterstroke, giving the impression that Presley cannot be contained on record and is literally jumping out at the buyer. The green and pink text, besides standing out against the photo, announces that *Elvis has landed*, and those two words are all that are needed. The album *is* Elvis Presley: energy, passion, and, in musical terms, it covers each of his stylistic influences bar gospel. Most important to remember is that, in 1956, teenagers were still buying singles, not long-playing albums. Put this on the shelves next to the Four Freshmen, Julie London and Ella Fitzgerald, all of whom released albums in 1956, and you can guess which one would get the youth vote.

RECORD LABELS

RCA

RELEASE DATES

No original UK release (US: March 1956)

SONGWRITERS

Howard Biggs/Joe Thomas, Bill Campbell, Ray Charles/Renald Richard, Dorothy LaBostrie/Little Richard, Rose Marie McCoy/Charles Singleton, Leon Payne, Carl Perkins, Don Robertson, Sydney Robin/Bob Shelton/Joe Shelton, Richard Rodgers/Lorenz Hart, Jesse Stone/Jimmy Wakely

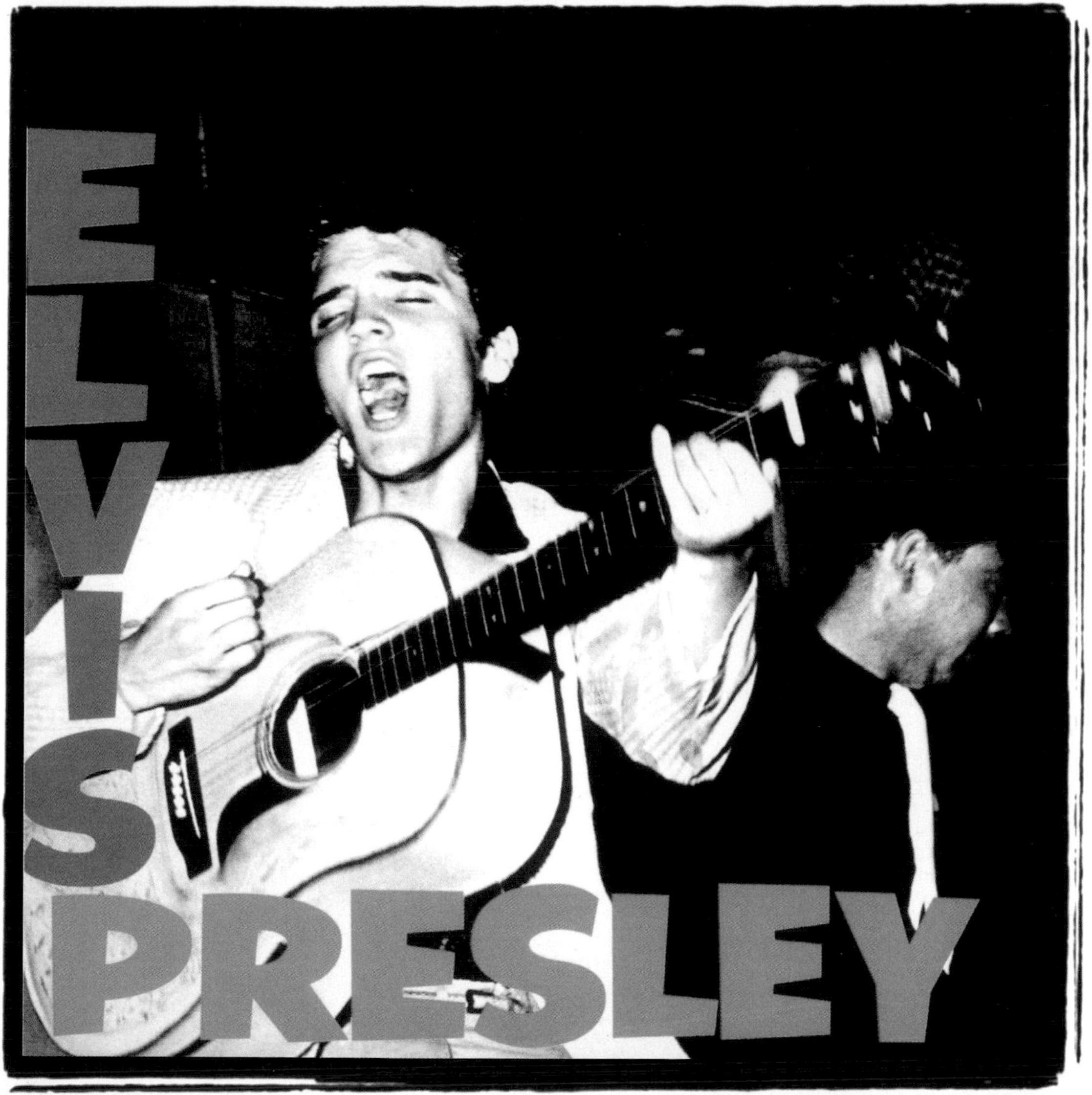
ELVIS
PRESLEY

PHIL WOODS & DONALD BYRD

THE YOUNG BLOODS (1956)

The title says it all: time for the old guard to move aside. Hard bop was the way forward for most young jazz players in the mid-50s: a more extreme version of bebop which, to the outsider, seemed only to pride itself on speed and how many notes and key changes could fit into a bar. Woods and Byrd were on alto sax and trumpet respectively, battling it out and weaving in among each other on this meeting of two bop greats. The yellow-and-red sleeve hints at an explosive meeting of minds: stand back, the touch paper has just been lit. This kind of stark, two-tone minimalism was commonplace for mid-50s bop sleeves, creating anticipation for the buyer, who would usually be looking for the most challenging work to listen to when it came to bop. The idea of playing one another off the stage (i.e. going head-to-head to see which musician played best) had been popular throughout jazz's entire history, but really came to the fore in the 40s and 50s, in the small clubs where the most cutting-edge jazz was played. Coupled with the colour scheme, the tiny image in the middle of the large sleeve perfectly puts across the feel of an intimate but volatile get-together.

RECORD LABELS

Prestige

RELEASE DATES

No original UK release (US: November 1956)

SONGWRITERS

Jimmy Davis/Roger Ramirez/Jimmy Sherman, Charlie Parker, Phil Wood

THE YOUNG BLOODS

PRESTIGE HI-FI   LP 7080

PHIL **woods** / DONALD **byrd**

HYDE/HANNAN

ELLA FITZGERALD

ELLA AND LOUIS (1956)

Jazz, as Ella Fitzgerald and Louis Armstrong knew it, was dead in 1956. If rock'n'roll hadn't swept all traces of the genre away entirely – and certainly hadn't with the adult record-buying public – Miles Davis was beginning to turn it inside out, forming a new 'cool', soon-to-be modal, jazz, the pinnacle of which would be **Kind Of Blue**, released in 1959. Not only that, but bebop had mangled the music the decade before and had by now become hard bop, making Ella and Louis two generations removed from what was really going on. For those that had aged with them, however, this sleeve promised the status quo: two old hands – two geniuses of the genre, in fact – getting together to play something relaxed and unchallenging, which would wrap you up warm in takes on old standards such as 'Moonlight In Vermont' or 'April In Paris'. No shades, no cigarettes: the down-homey sleeve would appeal to the right buyer and, against the odds, plenty of old rock'n'rollers would later be caught having similar photos of themselves taken for albums released when they were past their prime. Far from being an embarrassment, then, **Ella And Louis** sets the bar for growing old gracefully.

RECORD LABELS

Verve

RELEASE DATES

No original UK release (US: 1956)

SONGWRITERS

Irving Berlin, John Blackburn/Karl Sussedorf, Hoagy Carmichael/Ned Washington, Vernon Duke/'Yip' Harburg, George Gershwin/Ira Gershwin, Walker Gross/Jack Lawrence, Paul James/Kay Swift, Jerry Livingston/Al Neiburg/Marty Symes, Mitchell Parish/Frank Perkins

QUINCY JONES
THIS IS HOW I FEEL ABOUT JAZZ (1956)

While many modern jazz sleeves presented an elitist music, demanding that its listener be as enigmatic as the musicians looked, Quincy Jones, like Dizzy Gillespie, was very much an accessible, public relations-aware jazz musician. **This Is How I Feel About Jazz** is practically begging – not challenging – you to buy it. The bright, inviting colours give it a cool, modern edge, while the fact that Quincy is deigning to look out at the buyer is half the battle won for those who felt that jazz was beginning to push them away through sheer inaccessibility. Perhaps most importantly, however, Jones is holding a sheaf of yellowed papers, presumably scored music. Much mid-50s jazz, apart from the likes of Gil Evans' experiments with orchestration, was still built on pure improvisation, and so no cutting-edge jazz musician would be caught dead reading notated music. Here Quincy is harking back to big-band jazz imagery, where the likes of Duke Ellington would have full orchestras playing note-for-note reproductions of what their bandleader had composed. Jones was reminding buyers of jazz's rich history, while placing himself in it, aware of its lineage. Production for Michael Jackson aside, such awareness of things other than himself is probably one of the reasons why he is still so well known today.

RECORD LABELS

Mercury

RELEASE DATES

No original UK release (US: 1956)

SONGWRITERS

Principal songwriter: Quincy Jones

Secondary songwriters: Cannonball Adderly/Jon Hendricks, Harold Arlen/Truman Capote, Richard Carpenter

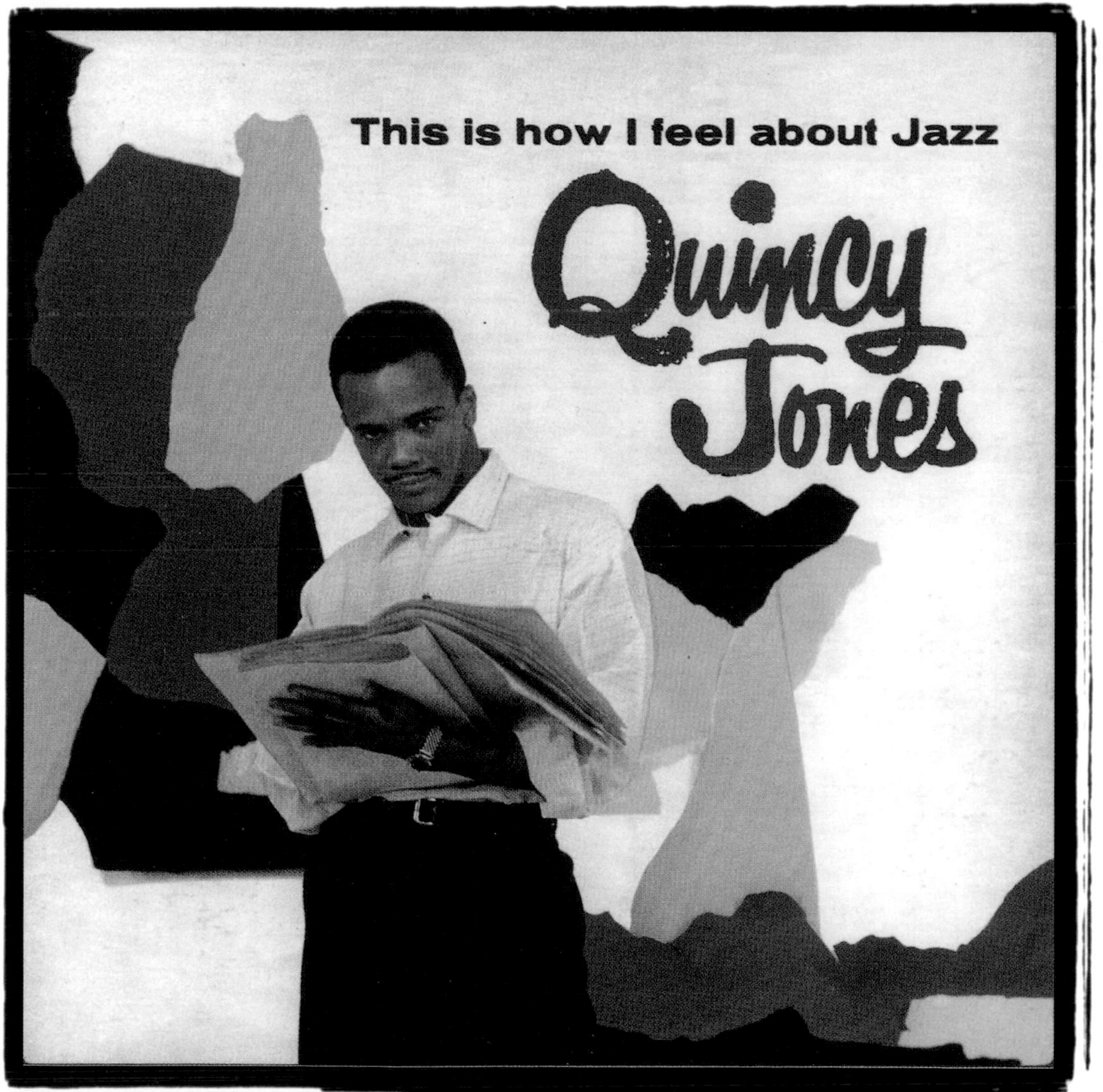
This is how I feel about Jazz
Quincy Jones

GERRY MULLIGAN & THELONIOUS MONK

MULLIGAN MEETS MONK (1957)

In jazz, unless you were Charlie Parker, Dizzy Gillespie or Miles Davis, chances were you would not be too instantly recognizable on the streets in the late 50s. Nevertheless, once you saw him, Thelonious Monk wasn't easily forgotten, being renowned for wearing outlandish hats; on this sleeve, however, he is rather sedately dressed in a flat cap. Unlike Ella Fitzgerald and Louis Armstrong, who by the 50s had been household names and faces for decades, Mulligan and Monk were probably more recognizable for their names – Thelonious Monk, certainly. The image is a concession here to the big, simple text trumpeting 'Mulligan Meets Monk'. The alliteration makes it easy to scan and the white jazz fan would know Gerry Mulligan's name right away, as one of the most famous white jazz players at the time. That he was meeting this enigmatic Monk character would be enticing, while the green-grey-and-red colour scheme of the text gives the record an almost European feel, perhaps adding to the promise of a sophisticated meeting of talents, pitched somewhere in between Ella & Louis and Phil Woods & Donald Byrd. Relying on the power of a name, Mulligan actually largely falls in on Monk's repertoire on this disc, as opposed to leading the session himself.

RECORD LABELS

Riverside

RELEASE DATES

No original UK release (US: August 1957)

SONGWRITERS

Gus Arnheim/Jules LeMare/Harry Tobias, Bernie Hanighen/Cootie Williams, Thelonious Monk, Gerry Mulligan

NEW YORK CITY, AUGUST, 1957   THELONIOUS MONK, GERRY MULLIGAN, WILBUR WARE, SHADOW WILSON

mulligan meets monk

ORIGINAL *Jazz* CLASSICS ®

FRANK SINATRA

COME FLY WITH ME (1958)

Here's Frank the reliable travel buddy. Or is it? Hold this up next to the solemn **In The Wee Small Hours Of The Morning** (1955) or the dangerously freewheeling **Songs For Swingin' Lovers** (1956), and **Come Fly With Me** looks like a caper waiting to happen. For his first album that is not about hard drinking or hard loving, Frank's collection of travel songs needed a bright and breezy image to signify the change in subject matter. But whose hand is he holding in the bottom left-hand corner? And why is he looking out at us, not her? Seems that Frank may still be a cad deep down, so don't be surprised if he jettisons you in Rio and shacks up with another party girl. Beautifully evocative of a certain type of themed record that existed into the 70s, **Come Fly With Me** still promises a ride full of surprises, and since it also contains the fruits of Frank's first, eclectic collaboration with arranger Billy May, it delivers.

RECORD LABELS

Capitol

RELEASE DATES

No original UK release (US: 1958)

SONGWRITERS

Ary Barroso/Bob Russell, Sammy Cahn/Jimmy Van Heusen, Carroll Coates, Matt Dennis/Tom Adair, Vernon Duke/E.Y. Harburg, Will Grosz/Jimmy Kennedy, Leo Robin/Ralph Rainger, Oley Speaks/Rudyard Kipling, Karl Suessdorf/John Blackburn, Victor Young/Harold Adamson

THE CECIL TAYLOR QUARTET

LOOKING AHEAD! (1958)

Much like Ornette Coleman's **The Shape Of Jazz To Come** sleeve from the following year, **Looking Ahead!** is pregnant with a talent waiting to explode. Coleman stares out at the buyer from his sleeve, but Taylor is too cool for that. He knows that his time has almost come and, with free jazz just around the corner in the early 60s, Taylor's sleeve and music are remarkably prescient. Looking away from the buyer and out to the future, a cigarette in one hand and his other arm stretched out on his piano, **Looking Ahead!** can be read two ways. Perhaps he is having a rest and a little smoke while waiting for the rest of jazz to catch up with him. Perhaps he is saying 'Try me.' It is almost too calm, as if he might jump up at any minute and take on anyone who tries to out-play him. The photo sees Taylor almost fighting for space with his name and title, suggesting that he was now becoming big enough to sell records on his name alone – but not so big that he only needed one name like 'Mulligan' or 'Monk'. This sleeve represents an artist on the cusp of the big time.

RECORD LABELS

Contemporary

RELEASE DATES

No original UK release (US: 1958)

SONGWRITERS

Principal songwriter: Cecil Taylor

Secondary songwriter: Earl Griffith

CONTEMPORARY
RECORDS S7562

THE CECIL TAYLOR QUARTET looking ahead!

ORNETTE COLEMAN

CHANGE OF THE CENTURY (1959)

With **Change Of The Century** Ornette Coleman went in the opposite direction of his almost apologetic, eager to please **The Shape Of Jazz To Come** LP from earlier in the year. With its bold lettering, neatly framed picture and smiling, horn-hugging Coleman, **The Shape Of Jazz To Come** was a partially smug, partially comforting sleeve that seemed to say, 'I know this might be a little weird, but trust me.' The problem was, not many did, and Ornette's free jazz was roundly derided in 1959. **Change Of The Century**, then, dispensed with the crowd-pleasing and went to the mystique-enhancing sleeve as seen from the likes of Charlie Parker. Far from the fresh-faced boy on the front of **...Jazz To Come**, **Change...** presents a man who has been put through the mill, and who knows that his further musical explorations on this album will exacerbate that. A haggard, unshaven musician looks out from this sleeve, beyond the audience, as if he is seeing something they can't. The fact is that he was, and this is the look – very similar to Marvin Gaye's on **What's Goin' On** – of a man who knows that he might get attacked for following his vision, but has to do it regardless.

RECORD LABELS

Atlantic

RELEASE DATES

No original UK release (US: October 1959)

SONGWRITERS

Ornette Coleman

ornette coleman   change of the century

ELVIS PRESLEY

50,000,000 ELVIS FANS CAN'T BE WRONG (ELVIS' GOLD RECORDS, VOL. 2) (1959)

Inspiring cover art from the likes of Bon Jovi to the Fall, the art for **50,000,000 Elvis Fans Can't Be Wrong** rivals its **Elvis Presley** predecessor (see page 21) for influence and instant recognizability. Released in November 1959, it had to make its mark. Elvis had been inducted into the American army in March the previous year, and many feared that while he was drafted his King Of Rock'n'Roll throne would be usurped. In 1959, however, the image of Elvis in a gold lamé suit epitomized success and reminded everyone that he was still the biggest-selling rock and pop performer of the times. It is important to remember that this was the first time a second 'greatest hits' album had been released by a rock'n'roll singer, and the fact that it consists of some single A- and B-side material, recorded in a rushed late-1958 session arranged while Elvis was still in the army, makes the power of the likes of 'A Big Hunk O' Love' even more remarkable. As a piece of PR, this was an inspired release.

RECORD LABELS

RCA

RELEASE DATES

No original UK release (US: November 1959)

SONGWRITERS

Dave Bartholomew/Pearl King/Anita Steiman, Bert Carroll/Moody Russell, Luther Dixon and Clyde Otis, David Hill/Aaron Schroeder, Ivory Joe Hunter, Jerry Leiber/Mike Stoller, Rose Marie McCoy/Cliff Owens, Bix Reichner/Sid Wayne, Bill Trader, Sidney Wyche

LSP-2075 (e)

STEREO
Electronically Reprocessed

50,000,000 ELVIS FANS CAN'T BE WRONG

ELVIS' GOLD RECORDS — Volume 2

A FOOL SUCH AS I
I NEED YOUR LOVE TONIGHT
WEAR MY RING AROUND YOUR NECK
DONCHA' THINK IT'S TIME
I BEG OF YOU
A BIG HUNK O' LOVE
DON'T
MY WISH CAME TRUE
ONE NIGHT
I GOT STUNG

MAGIC MILLIONS
RCA VICTOR
A "New Orthophonic" High Fidelity Recording

© RCA Printed in U.S.A.

ALBUM COVERS

RIFF HISTORY

SIXTIES

BOB DYLAN

BRINGING IT ALL BACK HOME (1965)

'Bringing It All Back Home' trumpeted the sleeve, but for many Bob Dylan followers, he was doing anything but on his first commercially released LP featuring an electric backing band. To many who didn't know that Dylan had grown up playing in rock'n'roll bands, he was leaving his folk roots behind by taking his music in this 'new' direction. Far from being the fresh-faced acoustic troubadour of his previous four albums, Dylan stares out from this sleeve, as photographed by Daniel Kramer, daring you to challenge his decision to, as many saw it, 'sell out' to the rock and pop market. With Sally Grossman, the wife of Dylan's then-manager Albert Grossman, sitting in the background, Dylan is surrounded by his influences, including albums by early folk mentor Eric Von Schmidt and blues legend Robert Johnson. A Lord Buckley LP rests on the mantelpiece as a signifier of Dylan's fascination with wordplay, while Dylan's own previous LP, **Another Side Of Bob Dylan** (1964), sits far behind him in the fireplace – part of the past, definitely not the future. **Bringing It All Back Home** is also credited for being the first Columbia Records LP not to feature the tracklisting on the front cover, breaking tradition with the favoured industry practice.

RECORD LABELS

CBS, Columbia

RELEASE DATES

May 1965 (US: April 1965)

SONGWRITERS

Bob Dylan

Bob Dylan
Bringing It All Back Home

THE BEATLES

RUBBER SOUL (1965)

Rubber Soul, the Beatles' second album released in 1965, could not have been more advanced than its predecessor **Help!** – not just musically, but visually too, as its artwork is not just another 'normal' shot of the boys together. The stretched effect of the sleeve's image came about by chance and is an example of how the group were always happy to use creative accidents to its advantage. The band were photographed at Lennon's house by Bob Freeman; when he was projecting the photo shoot results on a piece of 12-inch-by-12-inch cardboard, to illustrate what the images would look like on an LP sleeve, the piece of card was 'inadvertently tilted', Paul McCartney has recalled. 'The result was to stretch the perspective and elongate the faces…. The cover to our album **Rubber Soul** was decided.' The effect was also to present another new image of the group to the public, signifying the extraordinary rate at which the Beatles were maturing (**Help!** had only been released four months before), and suggests how keen they perhaps were to eradicate their smiley, mop-top image (see also the sleeves for **Sgt. Pepper's Lonely Hearts Club Band**, page 49 and **The Beatles**, page 63).

RECORD LABELS

Parlophone, Capitol

RELEASE DATES

December 1965

SONGWRITERS

George Harrison, John Lennon, Paul McCartney, Ringo Starr

THE 13TH FLOOR ELEVATORS

THE PSYCHEDELIC SOUNDS OF THE 13TH FLOOR ELEVATORS (1966)

It is hotly debated whether the 13th Floor Elevators invented psychedelic rock or not but, on the evidence of their debut album artwork, they were miles ahead of the game with its image. By 1967's 'Summer Of Love', this sort of thing would be commonplace, and stay that way up until the end of the decade in some circles. In 1966, however, this Elevators' sleeve was a sign of the coming times. Aside from its swirly, psychedelic font, the large eye, centre-sleeve, attracts the main focus, drawing upon later psych clichés such as the 'third eye' of Eastern mysticism, or even the 'mind's eye', something which, it was believed at the time, would only be truly opened with a healthy dose of LSD. The sleeve is a perfect indicator of the Elevators' mind-blowing psych-rock music, which included their only hit, 'You're Gonna Miss Me'. It was almost a case of too much too soon for the band, whose later album sleeves were more pared-down affairs, resulting in the almost pastoral artwork for their third and final album proper, **Bull Of The Woods** (1968).

RECORD LABELS

International Artists

RELEASE DATES

August 1966

SONGWRITERS

Roky Erikson, Tommy Hall, Powell St John, Stacy Sutherland

THE VELVET UNDERGROUND

THE VELVET UNDERGROUND AND NICO (1966)

Andy Warhol may have produced the Velvet Underground's debut album in name only, but his print artwork for the sleeve helped to secure the album's place in music history. The bright yellow banana on a plain white background perfectly captures the late-60s Pop Art movement as Andy Warhol saw it, using simple, recognizable popular culture imagery for mass production. Initial runs of the album had a 'Peel slowly and see' instruction next to the banana. On these, the banana skin was stuck on to the sleeve like a yellow skin, and buyers could peel it off to reveal a pinkish banana beneath. Producing these sleeves incurred extra costs that the record company, MGM, paid for, believing that Warhol's involvement with the record would boost sales. The back sleeve caused further controversy, being a live photo of the band with an upside-down image of Warhol actor Eric Emerson projected behind them. Emerson threatened to sue, so MGM had to recall the album from stores and airbrush his image out for further pressings. Although early sales of the album were disappointing, **The Velvet Underground And Nico** subsequently went down in history as being one of the most influential rock albums of all time.

RECORD LABELS

Verve

RELEASE DATES

October 1967 (US: December 1966)

SONGWRITERS

John Cale, Sterling Morrison, Lou Reed, Maureen Tucker

Andy Warhol

THE BEATLES

SGT. PEPPER'S LONELY HEARTS CLUB BAND (1967)

Not only is the **Sgt. Pepper...** sleeve a groundbreaking piece of art, designed by Peter Blake, but the packaging of the whole album alerted the world to what was truly possible within the confines of LP presentation. Though the Beatles' original idea of packaging the album with badges, pencils and other Pepper-related paraphernalia proved too costly, they provided an insert of Pepper cut-outs that included a Sgt. Pepper moustache, military 'wings', badges, a picture card of Sgt. Pepper and a cut-out stand of the band. For the sleeve itself, the group (in their guise as Sgt. Pepper's Lonely Hearts Club Band) decided to have themselves surrounded by famous faces they admired, as if they were being watched by a crowd of fans made up of their own heroes and other notables. It included Bob Dylan, Marlon Brando, Oscar Wilde and Madame Tussauds' own Beatles waxworks. Hitler, Jesus and Ghandi were requested but left out, while the Beatles had to personally write to Mae West to convince her to let them use her image. Everyone from *The Simpsons* to Frank Zappa has parodied the sleeve at one point or another, and it remains an iconic high point of British psychedelia and Pop Art.

RECORD LABELS

Parlophone, Capitol

RELEASE DATES

June 1967

SONGWRITERS

George Harrison, John Lennon, Paul McCartney

SCOTT WALKER

SCOTT (1967)

Never has one man been so determined to erase his past. Scott Walker found mega-fame in the mid-60s as one of the three-piece Walker Brothers, essentially the Take That of their day. Unimpressed with the screaming fans and adoration, however, Walker had become increasingly drawn to the seedy kitchen-sink dramas of Jacques Brel's songs and wanted to strike out on his own. Few were ready for the dark, often depressing nature of songs such as 'Amsterdam' and 'My Death'. Even less were ready for Scott, the once-grinning pop star, to emerge on an album sleeve looking like a man trying to escape the paparazzi. The scarf-and-shades look would define Walker for the rest of the decade – a European trapped in an American body – as he slowly withdrew from the popular music treadmill. More Greta Garbo than Robbie Williams, **Scott**'s sleeve is just the beginning of Walker the artist becoming more and more intangible across a series of album covers that has so far culminated in the brick wall-like sleeve of 2007's **And Who Shall Go To The Ball? And What Shall Go To The Ball?** Try breaking that barrier down.

RECORD LABELS

Philips, Smash

RELEASE DATES

August 1967

SONGWRITERS

Jacques Brel/Gérard Jouannest/Mort Shuman, Mack Gordan/Alfred Newman, Tim Hardin,Barry Mann/Cynthia Weil, André Previn/ Dorry Previn, Jack Segal/Robert Wells

SCOTT
SCOTT WALKER

PHILIPS

CREAM

DISRAELI GEARS (1967)

One of the greatest masterpieces of late-60s psychedelic artwork, **Disraeli Gears** reinvented Cream the blues rockers as Cream the psychedelic visionaries. LSD was a large part of the album's musical gestation, and that spilled out on to the album sleeve, which sees Ginger Baker, Jack Bruce and Eric Clapton almost entirely swamped by psychedelic swirls and blobs, with a large, winged 'CREAM' threatening to take flight, and take the record sleeve with it. The album's title is crammed in among the psychedelic imagery, almost looking like an afterthought. Perhaps this is because it was originally going to be entitled simply *Cream*, before (so the story goes) a roadie mispronounced 'derailleur gears' as 'Disraeli gears', inadvertently invoking the nineteenth-century politician Benjamin Disraeli. The artwork was designed by Australian psychedelic artist Martin Sharp, who had provided Eric Clapton with the lyrics to the album track 'Tales Of Brave Ulysses' on their first meeting.

RECORD LABELS

Reaction, Atco

RELEASE DATES

November 1967

SONGWRITERS

Principal songwriters: Ginger Baker, Pete Brown, Jack Bruce, Eric Clapton, Gail Collins, Felix Pappalardi

Secondary songwriters: Blind Joe Reynolds, Martin Sharp

THE DOORS

STRANGE DAYS (1967)

With their first album boasting a massive image of Jim Morrison's face on the sleeve, followed by tiny shots of band members Ray Manzerek, Robby Krieger and John Densmore, the Doors were determined never again to have an album so obviously promote them as 'Jim Morrison plus band'. With **Strange Days** they wanted to go in the opposite direction and Jim Morrison refused to appear on the cover at all. The original sleeve was going to be a shot of the band's image reflected in a mirror carried by a group of dwarves, but in the end a mock-up of a troupe of circus performers was shot in New York. In the photo, which wraps around on to the back sleeve, only the acrobats were real circus performers, with others in the group including friends, actors hired from agencies and even a taxi driver. In order to get around the fact that the band themselves weren't appearing on the sleeve, a poster advertising their previous album was pasted on the wall behind the strongman, ensuring that eagle-eyed buyers would know who had actually made the record, which otherwise didn't feature the band name or album title on the front.

RECORD LABELS

Elektra

RELEASE DATES

December 1967 (US: November 1967)

SONGWRITERS

John Densmore, Robby Krieger, Ray Manzerek, Jim Morrison

THE WHO

THE WHO SELL OUT (1968)

Demonstrating how album art can work on a number of different levels, **The Who Sell Out** was recorded and released at a time when the group actually were appearing in radio jingles advertising various consumer products. Conceived as something of an in-joke, the front sleeve sees Roger Daltrey in a bath of beans holding a giant tin of Heinz Baked Beans aloft, with the quote 'Thanks to Heinz Baked Beans everyday is a super day' beneath him. Pete Townshend, Keith Moon and John Entwistle are promoting their respective products, while the album itself was full of fake radio jingles poking fun at the band's own real careers as radio advertisers. The greatest irony, however, is that after the album was released, the Who were at the wrong end of a series of lawsuits from companies whose products were mentioned in the fake radio commercials and featured on the front and rear sleeves. Their third album, **The Who Sell Out** wasn't their best seller at the time either, so hopefully they had saved some royalties from the previous records to pay for this less-than-commercial interest.

RECORD LABELS

Track, Decca

RELEASE DATES

January 1968

SONGWRITERS

John Entwistle, John Keen, Pete Townshend

THE WHO SELL OUT

THE WHO SELL OUT

Replacing the stale smell of excess with
the sweet smell of success,
Peter Townshend, who, like nine out of ten stars,
needs it. Face the music with Odorono,
the all-day deodorant
that turns perspiration into inspiration.

This way to a cowboy's breakfast.
Daltry rides again. Thinks: "Thanks to Heinz
Baked Beans everyday is a super day".
Those who know how many beans make five
get Heinz beans inside and outside at
every opportunity. Get saucy.

THE SMALL FACES

OGDENS' NUT GONE FLAKE (1968)

The Small Faces were happy joining in with Britain's late-60s psych whimsy, but they also had a keen eye for nostalgia. The entire second side of **Ogdens'...** was given over to a fairy tale narrated by Stanley Unwin, a post-war comedian famous for his made-up language, 'Unwinese'. This fascination with old-world Britain permeated the artwork, which was a parody of the popular Victorian tobacco Ogdens' Nut Brown Flake. In this case, however, the Nut *Gone* Flake was surely designed as a reference to marijuana, which, when smoked, would get you 'gone', as the popular American slang term would have it. Original vinyl pressings of the album are now hard to find in a good enough condition to be enjoyed; they came in a gatefold sleeve that was cut out to look like a giant circular tobacco tin. Later CD pressings went so far as to put the CD in an actual tin case, though these are pretty easy to find, as they were being manufactured as recently as 2006.

RECORD LABELS

Immediate

RELEASE DATES

June 1968

SONGWRITERS

Kenney Jones, Ronnie Lane, Steve Marriott, Ian McLagan

BIG BROTHER & THE HOLDING COMPANY

CHEAP THRILLS (1968)

Columbia nixed the original idea for Big Brother & the Holding Company to call their second album *Sex, Dope And Cheap Thrills*, and so a simple **Cheap Thrills** it was. Legendary cartoonist and record collector Robert Crumb designed the iconic artwork after being asked by singer Janis Joplin to design a sleeve that looked like one of his then-popular *Zap Comix* strips. In Crumb's words, 'I was flattered, and I needed the money, so did it…. Of course they needed it, like, the next day, so I took some speed and worked all night on it.' The band didn't like his front sleeve, however, so used his rear sleeve design, which is why there are different 'windows' introducing the band, with one providing information on where the material was recorded. Most iconically, a comic strip-style Janis Joplin introduces a series of windows with: 'Playin' an' singing' fer yew the following tunes …'. Each song is then represented by an image, which is also written out for those who can't put two and two together. From the literal 'Ball And Chain' to the slightly more frivolous 'Combination Of The Two', Crumb's artwork opened the floodgates for him and he was still receiving requests to do album sleeves as late as 2005.

RECORD LABELS

CBS, Columbia

RELEASE DATES

September 1968 (US: August 1968)

SONGWRITERS

Principal songwriters: Peter Albin, Sam Andrew, David Getz, James Gurley, Janis Joplin

Secondary songwriters: Big Mama Thornton, Bert Berns/Jerry Ragovoy, George Gershwin/Ira Gershwin/DuBose Heyward

THE BEATLES

THE BEATLES (1968)

The Beatles has become better known as 'The White Album', thanks to its plain white sleeve featuring only 'The Beatles' embossed on the right-hand side, along with a stamped serial number. The stark artwork is in contrast to the sprawling 30-track double album of music the group released but, coming one year after the 'Summer Of Love', the album's artwork and its stripped-down music again sees the Beatles, along with Dylan and the Stones, pointing the way forward for a more sparse, roots-based music to take over in the late 60s/early 70s. According to longtime Beatles confidante Pete Shotton, the original idea behind the serial numbers was for one of them to be a 'winning number' in a lottery, until Shotton convinced McCartney that the Beatles didn't need that sort of gimmick to sell an album. Instead, the lowest numbers have become highly sought-after within collecting circles, though as the Beatles themselves were given the first four, it is unlikely that anyone will get their hands on those any time soon.

RECORD LABELS

Apple

RELEASE DATES

November 1968

SONGWRITERS

George Harrison, John Lennon, Paul McCartney, Ringo Starr

The BEATLES

SOFT MACHINE
THE SOFT MACHINE (1968)

Soft Machine came out of the late-60s UK Canterbury music scene and, along with the likes of Van Der Graff Generator, embodied the time when psychedelic rock was slowly turning itself into progressive rock. Pink Floyd would make that transition successfully between **The Piper At The Gates Of Dawn** and **The Dark Side Of The Moon**, but the likes of Soft Machine perhaps pointed the way to such transitions a little earlier. **The Soft Machine** was an ambitious mix of jazz and psych, with the group defining themselves via their unrivalled musical chops. Perhaps the prog came hand in hand with the mechanical names, as Soft Machine and VDGG moved away slightly from the whimsy of British psych darlings, proving themselves more on their musicianship. Their music demands more technical virtuosity – as opposed to simply 'feel' – which gets hinted at in the 'Machine' and 'Generator' parts of their names. The sleeve to Soft Machine's debut featured broken images of band members glimpsed in between the cogs of an unseen machine, presenting the group as a well-oiled, technically proficient group, with each member as important as the last when it came to making it work perfectly.

RECORD LABELS

Probe

RELEASE DATES

December 1968

SONGWRITERS

Kevin Ayers, Brian Hopper, Hugh Hopper, Michael Ratledge, Robert Wyatt

LED ZEPPELIN

LED ZEPPELIN (1969)

The Zeppelin has landed. Led Zeppelin were unique in rock'n'roll circles for having total artistic control over their music and album artworks. Anyone who has seen early Led Zeppelin press shots can imagine the horror of having a similarly hippified band photo emblazoned on the front of Led Zeppelin's debut album, as opposed to this iconic image, taken from a famous photo of the Hindenburg disaster. Hinting at the incendiary and epoch-making rock'n'roll inside, **Led Zeppelin**'s sleeve was a pen-and-ink drawing made by graphic designer George Hardie. Hardie's original concept was for the group to use the image of a Zeppelin airship flying in the clouds, as based on a San Franciscan club sign. Jimmy Page preferred the Hindenburg disaster shot, however, and used the rejected image on the rear sleeves of the first two Led Zeppelin albums. While wryly nodding to a comment from the Who's John Entwistle (or Keith Moon, depending on your source) that the group would 'go over like a lead balloon' (which gave Jimmy Page the name for the band when they were still known as the New Yardbirds), nothing says 'one of the most explosive albums of all time' like a burning airship.

RECORD LABELS

Atlantic

RELEASE DATES

March 1969 (US: February 1969)

SONGWRITERS

John Bonham, John Paul Jones, Jimmy Page, Robert Plant

IT'S A BEAUTIFUL DAY

IT'S A BEAUTIFUL DAY (1969)

The sky really was the limit for It's A Beautiful Day, whose incorporation of flute, violin and keyboards gave them a musical edge over some of their better-known San Francisco contemporaries. Though in tune with strands of the emerging prog-rock scene, they weren't afraid to incorporate folk, classical and, arguably, even world music into their wildly eclectic repertoire. While most of It's A Beautiful Day's West-Coast contemporaries were still working with designs that threatened a musical content designed to please Timothy Leary's 'turn on, tune in, drop out' crowd, **It's A Beautiful Day** reflects this album's musical diversity by showing much more scope than many other sleeves from the West Coast that year. Realizing that music didn't begin and end in LA or San Francisco, the band chose an album sleeve more in keeping with the nostaglic tones of some late-60s British acts (see **Ogdens' Nut Gone Flake**, page 58). Defiantly pre-war in its look, this is one sleeve that really does hint at the creative mix within, but makes it seem like a pleasant, breezy place to be – not something that can be said for some of the more intense works being produced by other West-Coast luminaries at the time.

RECORD LABELS

CBS, Columbia

RELEASE DATES

May 1969

SONGWRITERS

David LaFlamme, Linda LaFlamme, Vince Wallace

THE WHO

TOMMY (1969)

Nobody does heavy-handed better than Pete Townshend when he really wants to. Accompanying the double-album 'rock opera', which told the story of the eponymous deaf, dumb and blind pinball wizard, **Tommy** had a foldout triptych sleeve boasting a painting by Pop Artist Mike McInnerney. On the full artwork the large sphere, covered with diamond-shaped holes and taking up most of the middle panel, is partially sumberged with seagulls and clouds passing by. A large, spangly hand points towards the buyer from another panel: the idea of destiny and of Tommy being the Chosen One is also perhaps bound up in a 'Your country needs you'-style request, as Tommy asks his followers to blind, mute and deafen themselves in order to unlock their full spiritual potential. Baulking at a concept with no obvious link to the Who themselves, their record label, Track, demanded that the sleeve feature an image of the band. As a concession, small images of the band were put inside some of the diamond holes, but they were so tiny as to be almost unrecognizable.

RECORD LABELS

Track, Decca

RELEASE DATES

May 1969

SONGWRITERS

Principal songwriters: John Entwistle, Keith Moon, Pete Townshend

Secondary songwriters: Sonny Boy Williamson

TOMMY
the who

GRATEFUL DEAD

AOXOMOXOA (1969)

In the late 60s, particularly on America's West Coast, a flurry of now highly collectable posters were made to promote gigs in the popular clubs of the time. With fantastically intricate artwork, these posters reflected the psychedelic mindsets of those who attended the gigs and, of all the West-Coast bands, the Grateful Dead attracted the most psychedelically minded. It is no surprise, then, that their **Aoxomoxoa** sleeve is something of a watershed for the types of designs that would be found on these posters. Intricately detailed, and with the famous Dead skull'n'crossbones at the bottom, the artwork is bursting with creativity – and phallic symbolism. The design of the skull is unmistakably penis-shaped, and the testicle-like objects held in the skeletal hands further the case for the prosecution. Trees come bursting out of womb-like designs in the bottom corners, with rather suspect spermatozoa-shaped squiggles flowing into them. Getting one in under the censors, the sleeve only begins to hint at the further creative leaps that the Dead were making on their third album. One of the first albums to be recorded using 16-track tape machines, **Aoxomoxoa** proved there was life in the old psych dog yet, even as the Dead themselves would strip down **Workingman's Dead** later that same year.

RECORD LABELS

Warner Bros.

RELEASE DATES

October 1969 (US: June 1969)

SONGWRITERS

Jerry Garcia, Robert Hunter, Phil Lesh

ISAAC HAYES

HOT BUTTERED SOUL (1969)

As an album, **Hot Buttered Soul** broke the mould so much it was irreparable. Discarding all the rules about three-minute pop singles, it featured just four long tracks, with sweeping orchestration and spoken-word passages that influenced soul music throughout the 70s and hip hop in the 80s and 90s. Not only was the music innovative, but up until **Hot Buttered Soul**, soul connoisseurs were used to their album sleeves coming from the Motown school of design: grinning black artists such as Gladys Knight & the Pips or Smokey Robinson & the Miracles. Someone like Marvin Gaye might break the tradition by looking out dreamily at the buyer, but no one pushed their bald head up to the lens, making it impossible to see the singer's face. With **Hot Buttered Soul**, Hayes refused to toe the traditional line and made sure he didn't grin for the cameras. With his shades and heavy gold chain, he also created the look favoured by most hip-hop artists for ever more.

RECORD LABELS

Stax, Enterprise

RELEASE DATES

October 1969 (US: July 1969)

SONGWRITERS

Isaac Hayes, Al Isbell, Hal David/Burt Bacharach, Charles Chalmers/Charles Rhodes, Jimmy Webb

ISAAC
HAYES
Hot
Buttered
Soul

THE BEATLES

ABBEY ROAD (1969)

Of all of the Beatles' album covers, none has created such a stir as **Abbey Road**. When it was released in 1969 the public were already looking into the Beatles' music for hidden messages, and many had decided that they could hear Lennon saying 'I buried Paul' on **Sgt. Pepper's Lonely Hearts Club Band**. With **Abbey Road** the 'Paul Is Dead' rumours continued, mostly because he was crossing the zebra crossing outside the Abbey Road studios with bare feet and with his right leg forward – in contrast to the other, definitely living, Beatles' left legs being forward. With people reading it as a sign that Paul *was* now dead, Lennon, decked out in an all-white suit and heading the group, was seen as symbolizing a priest, Ringo (in a black suit and directly in front of Paul) an undertaker and George, in denims, bringing up the rear as gravedigger. Even more eerily, the number plate of the white Beatle car in the background reads 'LMW 281F', which is supposed to represent 'Linda McCartney, Widow' or 'Linda McCartney Weeps', with the '281F' signifying that Paul would have been 28, if he were alive when the photo was taken. Of course, it could all just be a coincidence. He would actually have still only been 27.

RECORD LABELS

Apple

RELEASE DATES

September 1969

SONGWRITERS

John Lennon, Paul McCartney, George Harrison, Ringo Starr

KING CRIMSON

IN THE COURT OF THE CRIMSON KING (1969)

As the birth of 70s prog rock as we know it, King Crimson's **In The Court Of The Crimson King** is almost a gateway into another world entirely, serving up mixed emotions: madness, fear, despair. Even Crimson's guitarist Robert Fripp has said, 'If you cover the smiling face, the eyes reveal an incredible sadness. What can one add? It reflects the music.' The cover was actually a painting by computer programmer Barry Godber, who died of a heart attack the year after **In The Court Of The Crimson King** was released. An illustration of the '21st Century Schizoid Man', it was Godber's one and only painting. While being one of the most arresting albums sleeves around, it is also probably one of the most off-putting. Listen too much and Crimson's music is likely to send you as mad as the Schizoid Man, but the music demands attention, and the open mouth almost acts as an 'enter if you dare' doorway. Madness is a favourite subject for prog-rock musicians and this album art promises it in spades.

RECORD LABELS

Island, Atlantic

RELEASE DATES

October 1969 (US: December 1969)

SONGWRITERS

Ian McDonald, Robert Fripp, Michael Giles, Greg Lake, Peter Sinfield

FRANK ZAPPA

HOT RATS (1969)

So in tune with the times was Zappa that, on his first solo album without the Mothers Of Invention, he was able to execute a work that perfectly captured the prevailing jazz-rock trend favoured by those wanting to ride the psych train to the end of the line. Furthermore, the sleeve for **Hot Rats** featured a number of late-60s touch points that captured the waning of an era. The prevailing psych movement was coming to an end in 1969, as the likes of Dylan, the Stones and the Beatles were going back to basics. Zappa's **Hot Rats** shot is something of a final epitaph for bright, psychedelic sleeves. Shot with an infrared camera, the image is visually striking, but the pink tone also suggests the sun going down for the last time. Miss Christine of the GTOs (Girls Together Outrageously, a Zappa-helmed girl group and world-famous collection of groupies), a late-60s celebrity within rock-star circles, looks set to crawl out of something remarkably grave-like. Her dark eyes have a distinctly zombified look and, all told, **Hot Rats** appears as something of a death knell for psychedelia.

RECORD LABELS

Reprise

RELEASE DATES

February 1970 (US: October 1969)

SONGWRITERS

Frank Zappa

FRANK ZAPPA

BIZARRE

RSLP

6356
STEREO

HOT RATS

PLASTIC ONO BAND

LIVE PEACE IN TORONTO 1969 (1969)

Though the only Plastic Ono Band album to be released, largely as a deterrent for bootleggers intending to trade the group's September 1969 show (on this date, the group included John Lennon, Eric Clapton, Klaus Voorman and Alan White – though it was more of a loose conceptual collection, rather than a fixed band), Lennon certainly made an effort on the artwork. Recorded just a month after the final **Abbey Road** sessions, the sleeve works on a number of levels with regards to Lennon's mindset. In his mind he was free of his Beatles obligations, so the clear blue sky represented the limitless bounds for whatever Lennon wanted to do (the ultimate realist, however, he had to let a cloud threaten to spoil the view). Also, Lennon was entering his 'bed-ins for peace' phase and working out ways of selling peace as a commodity. Everything he did at this time was centred around promoting the values of peace and love (hence the concert), and there was no more beautiful way of doing it than looking up to the heavens, forcing record buyers to look to their personal Utopias at the same time.

RECORD LABELS

Apple

RELEASE DATES

December 1969 (US: January 1970)

SONGWRITERS

Principal songwriters: John Lennon/Paul McCartney, Yoko Ono

Secondary songwriters: Janie Bradford/Berry Gordy, Carl Perkins, Larry Williams

ALBUM COVERS

SEVENTIES

MILES DAVIS

BITCHES BREW (1970)

With **Bitches Brew** Miles Davis began to consolidate his recent electronic experimentations. The artwork is a painting by German artist Mati Klarwein, and is in keeping with other out-there artworks of the time for rock musicians (see Santana's **Abraxas**, page 93). At this time Miles was courting a mixed audience, most of whom were rock and pop fans rather than jazz aficionados. He was also beginning to play in rock venues, exposing him to a much wider audience. The fact that he was breaking out, and even reaching out, is writ large on the **Bitches Brew** artwork. A Janus-like figure with one white face and one black face – the black staring out from the front of the sleeve, the white from the back – are joined together on the rear sleeve by a pair of entwined hands, again one white, one black. This fusion is brought to life in the music, which sets the template for all fusion jazz to come. From a jazz artist usually confined, in the wider public's mind, to a black audience, it was a mission statement like no other made before, and it showed Miles's older, more conservative jazz audience that he was still on the move.

RECORD LABELS

CBS, Columbia

RELEASE DATES

June 1970 (US: May 1970)

SONGWRITERS

Miles Davis, Wayne Shorter, Joe Zawinul

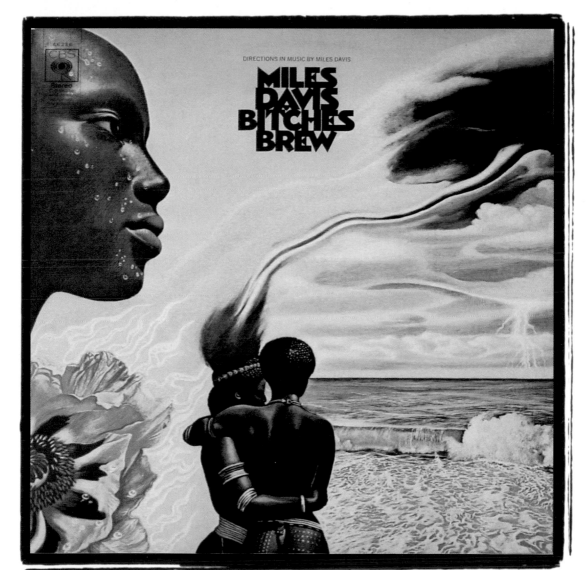

BOB DYLAN

SELF PORTRAIT (1970)

The album in which Bob Dylan tried to lose his fans. In 1969 he had become the first-ever bootlegged artist with **Great White Wonder**, a collection of unreleased studio outtakes. Lying low at the end of the 60s, Dylan thought if they wanted outtakes, he would give them outtakes, and hopefully they would leave him alone. However, loading up a double album with below-par live material, knock-off covers and a few lazy songs of his own backfired, and the clamour for the 'real' Bob Dylan to return grew stronger after he presented his fans with stuff they did not, in fact, want. **Self Portrait** flopped, but it was one big joke to Dylan anyway, who wasn't *really* saying this was the real him. 'There was no title for that album,' he said in 1984. 'I knew somebody who had some paints and a square canvas, and I did the cover up in about five minutes. And I said, "Well, I'm gonna call this album **Self Portrait**."' Hardly looking a thing like Dylan, in this case you *can* judge an album by its cover: knock-off, rushed, and an attempt by the artist to get people off his back.

RECORD LABELS

CBS, Columbia

RELEASE DATES

July 1970 (US: June 1970)

SONGWRITERS

Principal songwriter: Bob Dylan

Secondary songwriters: Gilbert Becaud/M. Curtis/Pierre Delanoe, Felice Bryant/Boudleaux Bryant, Alfrid Frank Beddoe, Paul Clayton/Larry Ehrlich/David Lazar/Tom Six, Lorenz Hart/Richard Rodgers, Gordon Lightfoot, John Lomax/Alan Lomax/Frank Warner, Cecil A. Null, Paul Simon

THE MOTHERS OF INVENTION

WEASELS RiPPED MY FLESH (1970)

Never one to miss an iconic image, Zappa's sleeve for **Weasels Ripped My Flesh** parodied the then well-known advertising campaign for Shick electric razors. Of course, this being Frank Zappa, a comfortable electric razor designed for the businessman on the go is replaced by an electrically powered musteline ripping the flesh from its user's face. The idea came from a 1956 issue of American men's adventure magazine *Man's Life*, the cover story of which was 'WEASELS RIPPED MY FLESH', featuring an illustration of a topless man being butchered by weasels in what appears to be a lake or river. Zappa took the magazine to illustrator Neon Park (who also designed many Little Feat album covers) and asked him, 'What can you do that's worse than this?'. It is not just that the weasel is tearing this man's flesh from his face, of course, but that the man is actually letting it. 'RZZZZZ!' the weasel-razor goes, giving its user the close shave of a lifetime. Oddly, the German sleeve for the album showed a tin baby caught in a rat-trap; the link is still unclear.

RECORD LABELS

Reprise

RELEASE DATES

September 1970

SONGWRITERS

Principal songwriter: Frank Zappa

Secondary songwriter: Little Richard

SANTANA

ABRAXAS (1970)

This is an example of how, as a fledgling musician, to make your album one of the most recognizable of all time, and ensure that an artist gets enough commissions to live off of the profits. Mati Klarwein had created the *Annunciation* painting in 1961 when he was just 28, and Carlos Santana was looking for something to adorn the sleeve of his second and early career-defining record. Abracadabra! (which the album title sometimes translates as) – he came across Klarwein's painting reproduced in a magazine and the sleeve was found. In Klarwein's surprised words, 'It did me a world of good. I saw the album pinned to the wall in a shaman's mud hut in Niger and inside a Rastafarian's ganja hauling truck in Jamaica. I was in global company'. Not only did the use of *Annunciation* help to launch two careers, however, but it perfectly encapsulated the fusion spirit of the early 70s, in which psychedelic imagery matured to take on almost religious overtones, as seen on Klarwein's art for Miles Davis (see **Bitches Brew**, page 87 and **Live-Evil** [1970]) and in the 1970 painting that was later used for the Last Poets' 1995 **Holy Terror** LP.

RECORD LABELS

CBS, Columbia

RELEASE DATES

November 1970 (US: September 1970)

SONGWRITERS

Principal songwriters: José Chepitó Areas, Mike Carabello, Albert Gianquinto, Carlos Santana, Gregg Rolie

Secondary songwriters: Peter Green/Gabor Szabo, Tito Puente

PINK FLOYD

ATOM HEART MOTHER (1970)

The early 70s seemed to be a time for longstanding acts to want to erase their established image and either court new audiences, or no audience at all. Dylan had done it with **Self Portrait** (see page 89), the Grateful Dead with **American Beauty** (see page 97) and by 1970 Pink Floyd were looking to be taken on purely musical terms, without the baggage of the public's preconceived notions. Along with Hawkwind and other bands that had come out of London's UFO Club scene, Pink Floyd were firmly entrenched in the public consciousness as space rockers. Wanting to eradicate this and let the music speak for itself, Pink Floyd sent their longtime artist Storm Thorgerson out to design a sleeve, with the instruction to make something plain. Inspired by Andy Warhol's *Cow Wallpaper*, Thorgerson promptly went out into the country and took a photo of a cow in a field; in doing so he created probably the weirdest, but definitely the most image-distorting album sleeve of the band's career, and also made an unwitting superstar out of the bovine, Lulubelle III.

RECORD LABELS

Harvest

RELEASE DATES

October 1970

SONGWRITERS

David Gilmour, Nick Mason, Roger Waters, Richard Wright

GRATEFUL DEAD

AMERICAN BEAUTY (1970)

When the Dead caught up with the prevailing country-rock trends of the late 60s and early 70s, they did it in such style as to make you think that they invented it. Everything about country rock suggested going back to the old homestead and as such **American Beauty** was light years behind the artwork for **Aoxomoxoa** (see page 73). But the group didn't waste the talents of their intricate sleeve designers Alton Kelley and Stanley Mouse. Even the wooden mount on which the ornate emblem stands was designed with such care and detail as to look like an actual piece of wood. The brass emblem was similarly detailed, while the delicate fragility of the single rose perfectly encapsulated how country-rock artists at the time strove for musical perfection and a deft simplicity; not for a couple of years would country rock become the bloated shadow of its former self. As a way of introducing a band's new direction, **American Beauty** is a perfect artwork. Far from the sonic experiments of earlier albums, and following on from their previous excursion into country rock, **Workingmans' Dead**, **American Beauty** (some say the album title reads 'American Reality') encapsulates the promise of a folk-, bluegrass- and country-influenced Grateful Dead album.

RECORD LABELS

Warner Bros.

RELEASE DATES

December 1970

SONGWRITERS

Jerry Garcia, Robert Hunter, Phil Lesh, Ron McKernan, Bob Weir

SEVENTIES

THE GRATEFUL DEAD

AMERICAN BEAUTY

THE ROLLING STONES

STICKY FINGERS (1971)

Not just notable for its innovative sleeve, **Sticky Fingers** was the first Rolling Stones album to feature the now ubiquitous Rolling Stones 'tongue and lips' logo in its artwork. A genius piece of band logoing, it raised the Stones up as one of the first rock'n'roll artists to actively market themselves as a product of sorts, while the logo acted as a quality-control stamp for a brand to be trusted – for the time being. The early 70s saw the Stones at their most depraved, and most of the songs on **Sticky Fingers** were overtly concerned with drug use. Why were their fingers sticky? Well, the front of the sleeve actually featured a working jeans fly which, when unzipped, revealed a man's white cotton pants. Maybe that had something to do with it. That the cover shot was photographed by notorious sleaze-hound Andy Warhol definitely did. **Sticky Fingers** stands out in a time where artists seemed to want to update their image for the new decade. Instead of re-branding themselves as nice boys, the Stones seemed keen to plough the ever-popular 'nasty boys of rock' furrow. Anything that would continue to cultivate their early 'hide your wives and lock up your daughters' image was fine by them.

RECORD LABELS

Rolling Stones

RELEASE DATES

April 1971

SONGWRITERS

Principal songwriters: Mick Jagger, Keith Richards

Secondary songwriter: Marianne Faithfull

THE WHO

WHO'S NEXT (1971)

The Who Sell Out (see page 57) might have been a bit of a laugh, and **Tommy** (see page 71) saw Townshend indulging himself, but the **Who's Next** sleeve trumpeted that this was a band returning to the no-nonsense rock that they did best, presenting an album book-ended by now-staple rock classics 'Baba O'Riley' and 'Won't Get Fooled Again'. And how did they announce their return to balls-out (quite literally) rock'n'roll? By choosing an album sleeve that pictured the group having just finished collectively urinating on a giant monolith. The monolith, of course, is most commonly linked with *2001: A Space Odyssey* (1968), a behemoth of a film that, at that time, still represented everything about the longest, most overblown extremes of film-making. So could this also be having a dig at Led Zeppelin, that rock monolith which had, in the past few years, become in some eyes the most overblown rock band of them all? Daltrey certainly has a Robert Plant-baiting look about him on this sleeve, and Townshend might be boasting that this band is symbolically pissing on all others who wish to grasp their rock crown.

RECORD LABELS

Track, Decca

RELEASE DATES

September 1971 (US: August 1971)

SONGWRITERS

Principal songwriter: Pete Townshend

Secondary songwriter: John Entwistle

SLY & THE FAMILY STONE

THERE'S A RIOT GOIN' ON (1971)

Just as **Hot Buttered Soul** (see page 75) revolutionized the sight and sound of soul in the late 60s, Sly Stone was doing it again in 1971 with the murky, lo-fi **There's A Riot Goin' On**. A pointed reference to Sly's own late arrivals or even no-shows at concerts – and the subsequent riots that sometimes ensued – there was even a 'number-less' title track with the running time 0.00, which, depending on who you believe, was either Sly's way of saying that it wasn't his fault, or that there shouldn't be any riots. The sleeve featured a red, white and black American flag with suns in place of the stars, mocking the patriotism of a country still at war with Vietnam. Sly actually had a real flag with this design made and photographed; he described the sleeve as representing 'people of all colours', with the red uniting the colours and signifying blood, the one thing that everybody has in common. Sly explained changing the stars to suns by saying, '[The designer of the American flag] Betsy Ross did the best she could with what she had. I thought I could do better.' The lack of a band name or album title shows just how big Sly & the Family Stone were at the time of this album's release.

RECORD LABELS

Epic

RELEASE DATES

January 1972 (US: November 1971)

SONGWRITERS

Sylvester Stewart

TRAFFIC

THE LOW SPARK OF HIGH HEELED BOYS (1971)

Traffic was the archetypal hard-rock band of the early 70s, and the artwork for **The Low Spark Of High Heeled Boys** perfectly encapsulates that time, when British rock was splintering off down prog-, jazz- and blues-rock avenues. The members of Traffic were talented enough to straddle it all to a degree; this sleeve in particular features all the visual hallmarks of a 70s British rock band with proven chops. Two elements, air and earth, appear on two sides of the cube, imagery beloved of rock bands (while water also features heavily in the lyrics). The chessboard reflects the title track, a 12-minute musing on the music business, the game that Traffic found themselves caught up in. It is simple imagery and you can make the connections without too much strain. The clinical look of the sleeve, however, is what is most important here. Less interested in feel and experimentation than they were in technical proficiency, bands such as Traffic and other early 70s hard-rock heroes such as Gary Moore would pride themselves on their slick production and guitar histrionics. It is the kind of album sleeve that sets its stall out clearly: smooth, well-recorded rock, hints of prog, touches of jazz, but nothing to completely blow your mind.

RECORD LABELS

Island

RELEASE DATES

December 1971

SONGWRITERS

Principal songwriters: Steve Winwood, Jim Capaldi

Secondary songwriters: Ric Grech/Jim Gordon, Anna Capaldi

THE ALLMAN BROTHERS BAND

EAT A PEACH (1972)

One of the greatest Southern-rock bands of all time, the Allman Brothers Band were at the height of their collective powers when they recorded **Eat A Peach**. At a time when West-Coast rock was being critically hailed and reaping the most commercial rewards, the Allman Brothers Band were looking to celebrate their southern roots, so **Eat A Peach**'s sleeve was drawn up: a giant peach being dragged along on a huge, flatbed truck. The album's original title was going to be *The Kind We Grow In Dixie*, suggesting that the music inside was as juicy, succulent and impressive as the peach in the image. The band weren't happy with the name, however, and in the end went with **Eat A Peach**, after Duane Allman's idiosyncratic views on Vietnam were quoted in an interview: 'There ain't no revolution, it's evolution, but every time I'm in Georgia I eat a peach for peace, the two-legged Georgia variety.'

RECORD LABELS

Capricorn

RELEASE DATES

February 1972

SONGWRITERS

Principal songwriters: Duane Allman, Gregg Allman, Dicky Betts

Secondary songwriters: Steve Alaimo, Donovan, Elmore James/Marshall Sehorn/Sonny Boy Williamson, Jai Johanny Johanson, Berry Oakley, Butch Trucks, Muddy Waters

LITTLE FEAT

SAILIN' SHOES (1972)

Bucking against the trend of somewhat overwrought and elaborate album artwork that characterized the late-60s psychedelic explosion, the early 70s saw a run of 'does what it says on the tin' album sleeves coming from back-to-basics rock bands. **Eat A Peach** (see page 107) offers said fruit for the taking, while **Sailin' Shoes** sees a pair of shoes sailing off from a cake on a swing. Neon Park designed the sleeve. He had previously worked on Frank Zappa's **Weasels Ripped My Flesh** (see page 91), and brought that oddball humour to **Sailin' Shoes**, which is regarded as a take on Jean-Honoré Fragonard's *The Swing*. Instead of a beau reaching up from the flowers towards his swinging lover, however, a rather bug-eyed snail is getting a right old eyeful up this cake's skirt. Park's distinctive art would feature on almost every Little Feat sleeve from hereon in, and would at times spoof the Stones' 'tongue and lips' logo. (Incidentally, the figure in the background is apparently meant to be Mick Jagger dressed as Thomas Gainsborough's *Blue Boy*.)

RECORD LABELS
Warner Bros.

RELEASE DATES
May 1972

SONGWRITERS
Principal songwriters: Lowell George, Bill Payne

Secondary songwriters: Richard Hayward, Martin Kibbee

Little Feat / Sailin' Shoes

THE ROLLING STONES
EXILE ON MAIN STREET (1972)

By the time the Stones were ready to record **Exile On Main Street**, they had been forced to leave England as tax exiles and were holed up in Villefrance-sur-Mer, near Nice in the South of France. Keith Richards had hired Nellcôte, a mansion on the waterfront, believed to be the Gestapo headquarters during the Second World War, and it was from there that the Stones' masterpiece was recorded. The double album's sleeve art reflects the drug-ridden, carnivalesque troupe that the band had become by the early 70s and, in designer John Van Hamersvled's words, the tone was 'anarchy – drug dealers and freaks and crazy people left over from the 60s'. Freak show 'attractions' and vaudevillian entertainers all populate the sleeve (apparently a photograph taken of a tattoo parlour wall on Route 66), capturing the somewhat chaotic spirit of the recording sessions, which saw Keith Richards dealing with heroin addiction, the continual presence of friends and hangers-on, and recording sessions that wouldn't often see the band actually recording together. The images on the sleeve might say 'freaks', but the actual art design captures the sprawling, rough-edged recording, right down to Mick Jagger's handwritten title.

RECORD LABELS
Rolling Stones

RELEASE DATES
June 1972

SONGWRITERS
Principal songwriters: Mick Jagger/Keith Richards

Secondary songwriters: Slim Harpo, Robert Johnson, Mick Taylor

ALICE COOPER

SCHOOL'S OUT (1972)

'Shock-rocker' Alice Cooper was about to make it huge in the heavy-metal mainstream with this album, but the artwork (and, indeed, the music) showed that he was thinking far beyond the school of thought which believed that all you needed to sell a rock album was a scantily clad girl, preferably in no small degree of danger (see Whitesnake's **Love Hunter** [1979]). A smart design, Cooper's **School's Out** sleeve reminds everyone of the excitement they felt when school finished for the summer, thereby suggesting that they could relive that experience if they would just buy his album. A lovely detail sees all five band members' initials (and a couple of surnames) etched into the school desk design, which actually opened up as an American desk would. As a concession to the tits'n'ass end of the rock market, the record itself came wrapped in a pair of white, blue or pink knickers. Mint condition copies of this sleeve, as long as they include the knickers, can now change hands for up to £75.

RECORD LABELS

Warner Bros.

RELEASE DATES

July 1972 (US: June 1972)

SONGWRITERS

Michael Bruce, Glen Buxton, Alice Cooper, Dennis Dunaway, Neal Smith

ROD STEWART

NEVER A DULL MOMENT (1972)

The great British sense of humour wins out on Rod Stewart's **Never A Dull Moment**, and it is probably not far from the sort of wry humour that Stewart and cohort Ron Wood would reel off with each other in the recording studio. At the time of recording **Never A Dull Moment** Stewart's solo career was really taking off, but at the same time he still fronted the Faces, a situation that led to some tensions in the band. **Never A Dull Moment** sees Stewart trying to come to terms with his new-found fame. On the autobiographical opener 'True Blue' he decides he would really rather be back home, and the sleeve pictures him there, sitting in an armchair, boxed in by the decidedly non-rock'n'roll beiges, bottle-greens and mustard-yellows of a typical 70s front room. Though Stewart still manages to look as cool as you like, the expression on his face says it all. Never a dull moment? More like ain't nothing happening round here, mate. And with Stewart arguably having reached his peak with this album and its predecessor, **Every Picture Tells A Story** (1971), he never had a better moment than he did in the early 70s.

RECORD LABELS

Mercury

RELEASE DATES

July 1972

SONGWRITERS

Principal songwriters: Rod Stewart, Ron Wood

Secondary songwriters: Sam Cooke, Bob Dylan, Bill Foster/Ellington Jordan, Jimi Hendrix, Martin Quittenton

ROD STEWART
NEVER A DULL MOMENT

mercury

YES

CLOSE TO THE EDGE (1972)

Often cited as the greatest prog-rock album of all time, Yes's **Close To The Edge** is, visually speaking, one of the most modest. Compare it to their earlier albums – most showing dull band shots that presented the less exciting end of 70s fashion – it stands out for its innovative design; compare it to their later ones, and it is notable for its subtlety. Later Yes albums played fully into the prog stereotype of convoluted sleeve art, reflecting the deeper message most artists thought they were conveying. With its simple green sleeve, however, **Close To The Edge** suggests some degree of spiritual cleansing. The title song is said to have been influenced by Herman Hesse's *Siddhartha*, marking the eponymous character's tracks to the edge of a river (again chiming with the green sleeve), where he finds spiritual awakening. Interestingly, the album also reproduces the 'Yes' logo as seen on the previous year's **Fragile**, suggesting that the group were beginning to find an identifiable look. Prog may have been derided in some circles, but in the early 70s Yes were a commercial force, and relying on a band logo above anything else is something that only bands such as the Stones would have been able to enjoy up to this point.

RECORD LABELS

Atlantic

RELEASE DATES

September 1972

SONGWRITERS

Jon Anderson, Bill Bruford, Steve Howe, Rick Wakeman

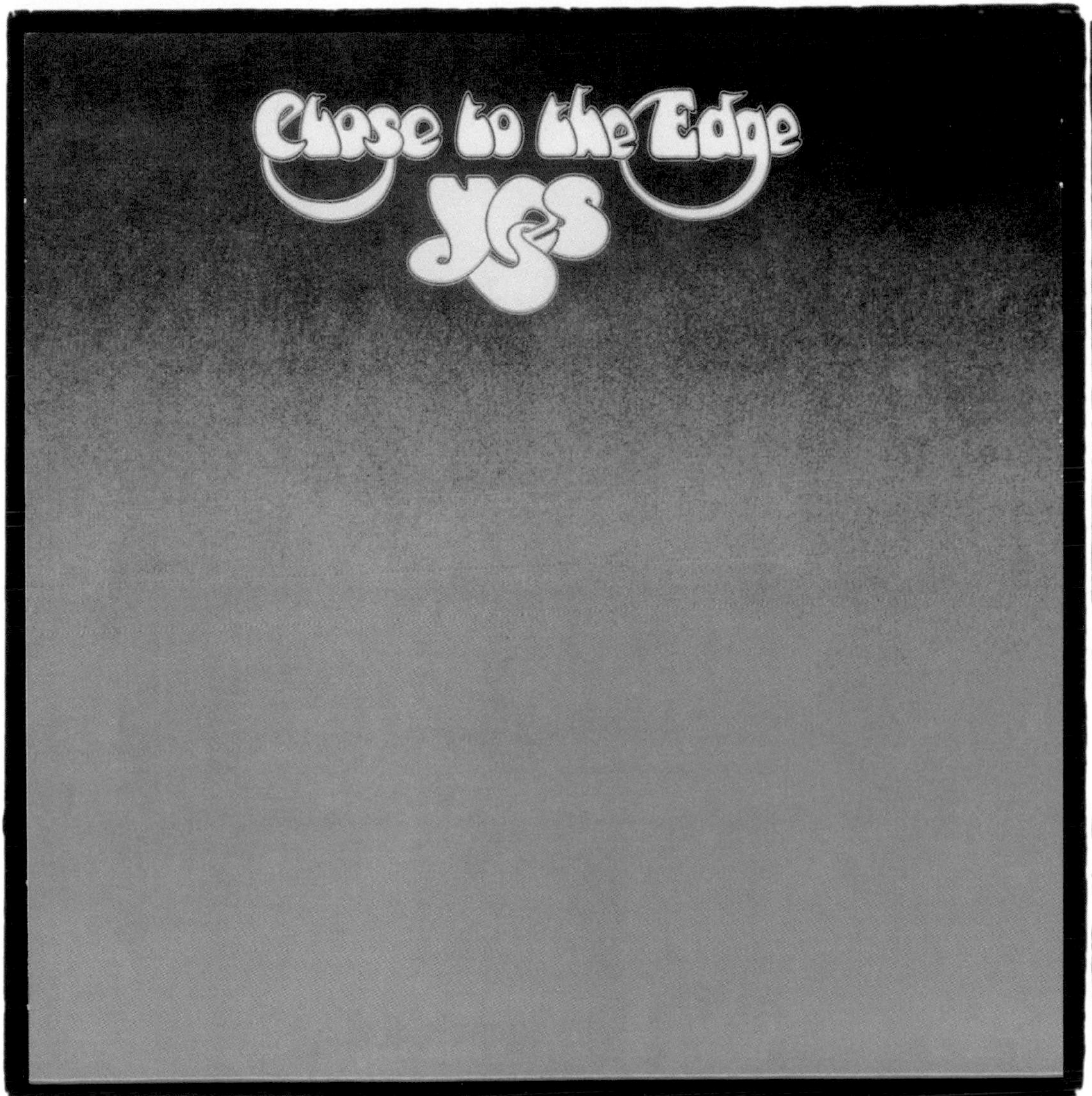
Close to the Edge
Yes

PINK FLOYD

THE DARK SIDE OF THE MOON (1973)

Pink Floyd's light shows had been a hallmark of the band's live performances since their 60s days performing in London's UFO Club. Storm Thorgerson and Aubrey Powell reflected this with one of the most iconic album sleeves of all time for **The Dark Side Of The Moon**, with their design of a prism refracting white light into a spectrum: simple physics and something record buyers might relate to from school. 'The prism was a way to talk about the fact that this band … would do light and sound [at their shows],' Thorgerson has recalled. It is also claimed that Roger Waters wanted a sleeve that reflected 'madness of ambition', and that the triangle is the symbol for ambition. The spectrum runs throughout the inner sleeve, tracing a heartbeat's sonar wave. Pink Floyd also gave away two posters and a set of stickers with the packaging, one of the posters being a montage shot of the band playing live, the other being a photo of the Great Pyramids of Giza taken on infrared film (the stickers were also of the pyramids).

RECORD LABELS

Harvest

RELEASE DATES

March 1973

SONGWRITERS

David Gilmour, Nick Mason, Roger Waters, Richard Wright

THE FACES

OOH LA LA (1973)

Although, thanks to Rod Stewart's monumental solo success, the Faces were in some state of dissolution when they recorded this, their final album, at least they still managed to create a witty sleeve that stands the test of time. While searching for sleeve ideas, designer Jim Ladwig uncovered an old 1930s toothpaste advert featuring an image of pre-war radio comedian Fred Allen. It had a tab that, when pulled, moved Allen's eyes and mouth in a comical manner, which gave Ladwig an idea. He mocked up an LP-sized copy of the sleeve and showed it to Ron Wood, who replied, 'Ooh la la!'. Taking that as the album's title, the group remade the advert with a French man-about-town where Fred Allen had once been, and a shot of the Faces getting an eyeful up the skirt of a can-can dancer on the rear sleeve. This, too, had its interactive joys, as the dancer could be made to kick when another tab was manipulated. The concept was so catchy that the Faces' record label produced scaled-down replicas to send to journalists.

RECORD LABELS

Warner Bros.

RELEASE DATES

April 1973

SONGWRITERS

Kenney Jones, Ronnie Lane, Ian McLagan, Rod Stewart, Ron Wood

OOH LA LA
faces

BOB MARLEY & THE WAILERS

CATCH A FIRE (1973)

Island label founder Chris Blackwell had such faith in Bob Marley & the Wailers that he wanted their major label debut to make a real impression. The sleeve that designers Rod Dyer and Bob Weiner came up with was a beauty, but impractical. Designed to look like a Zippo lighter, the sleeve opened up like a clapperboard, with a hinge on the side keeping both halves of the cover together. The record sat in the bottom pouch so that, when you opened the sleeve, you could see the top of it poking out like the flint on a proper Zippo. It was a great idea, and marijuana-smokers would have easily caught on to the reference. Sadly it proved difficult for Island to manufacture, as no machine could fix the top and bottom halves together properly. Instead, each sleeve had to be handmade, so only the first run of 20,000 actually ended up using this innovative design. From that point onwards, the standard 'Bob Marley with a joint' sleeve was designed and used, showing Bob doing what most were with their real Zippos, and leaving the original copies of the album now worth £40 to collectors.

RECORD LABELS

Island

RELEASE DATES

April 1973

SONGWRITERS

Bob Marley, Peter Tosh

SPRP 9241

The Wailers
Catch A Fire

LED ZEPPELIN

HOUSES OF THE HOLY (1973)

Longtime Pink Floyd sleeve designer Storm Thorgerson had originally created artwork for **Houses Of The Holy**, which was simply a bright tennis court with a tennis racket on it. Not a band to take criticism lightly at the time, Led Zep sacked Thorgerson for metaphorically suggesting that their music was a racket, and asked for someone else from the noted British design company, Hipgnosis, to come up with a sleeve. The result was striking, largely thanks to a happy accident and a painstaking collage made up of 30 photos. Hipgnosis designer Aubrey Powell led a photoshoot in rainy, cloudy conditions for an entire week, trying to capture the perfect sunrise at the Giant's Causeway in Northern Ireland. Not much was forthcoming without the right weather, but what Powell did capture was tinted by accident, which resulted in the children getting a purple hue. Unwittingly, Powell's sleeve chimed perfectly with its influence, Arthur C. Clarke's novel *Childhood's End*, which saw a barely human, child-like race being all that remained from a group of children quarantined on a continent. The children had morphed with the Overmind, an intangible being of pure energy, something Led Zeppelin pretty much resembled in 1973.

RECORD LABELS

Atlantic

RELEASE DATES

April 1973

SONGWRITERS

John Bonham, John Paul Jones, Jimmy Page, Robert Plant

DAVID BOWIE

ALADDIN SANE (1973)

'Ziggy Stardust goes to America' was Bowie's description of **Aladdin Sane**, a collection of songs he mostly wrote while touring the States in 1972. Compare this sleeve with the career-making **The Rise And Fall Of Ziggy Stardust And The Spiders From Mars** (1972) and you can see that Ziggy, who did not originally have the magenta mullet with which he became synonymous, is a more stylish, Americanized character than he was on the front of the much more English-looking **Ziggy Stardust…** sleeve. The slick, carefully designed and self-aware portrait marked the pinnacle of glam, and was an image that would look fantastic on American billboards. It also came to define Bowie in most people's minds: the pale skin, bright red mullet and lighting bolt across the face are impossible to confuse with anyone else in the history of music. For Bowie, however, it was just one more image change. Though he kept the mullet for **Diamond Dogs** (see page 139), that sleeve again saw an evolution, with Bowie appearing much more ghostly as glam died out. In just a few years' time Bowie would adopt a shorter side-parting, as the last vestiges of Ziggy were left behind and Bowie morphed into the Thin White Duke.

RECORD LABELS

RCA

RELEASE DATES

April 1973 (US: May 1973)

SONGWRITERS

Principal songwriter: David Bowie

Secondary songwriters: Mick Jagger/Keith Richards

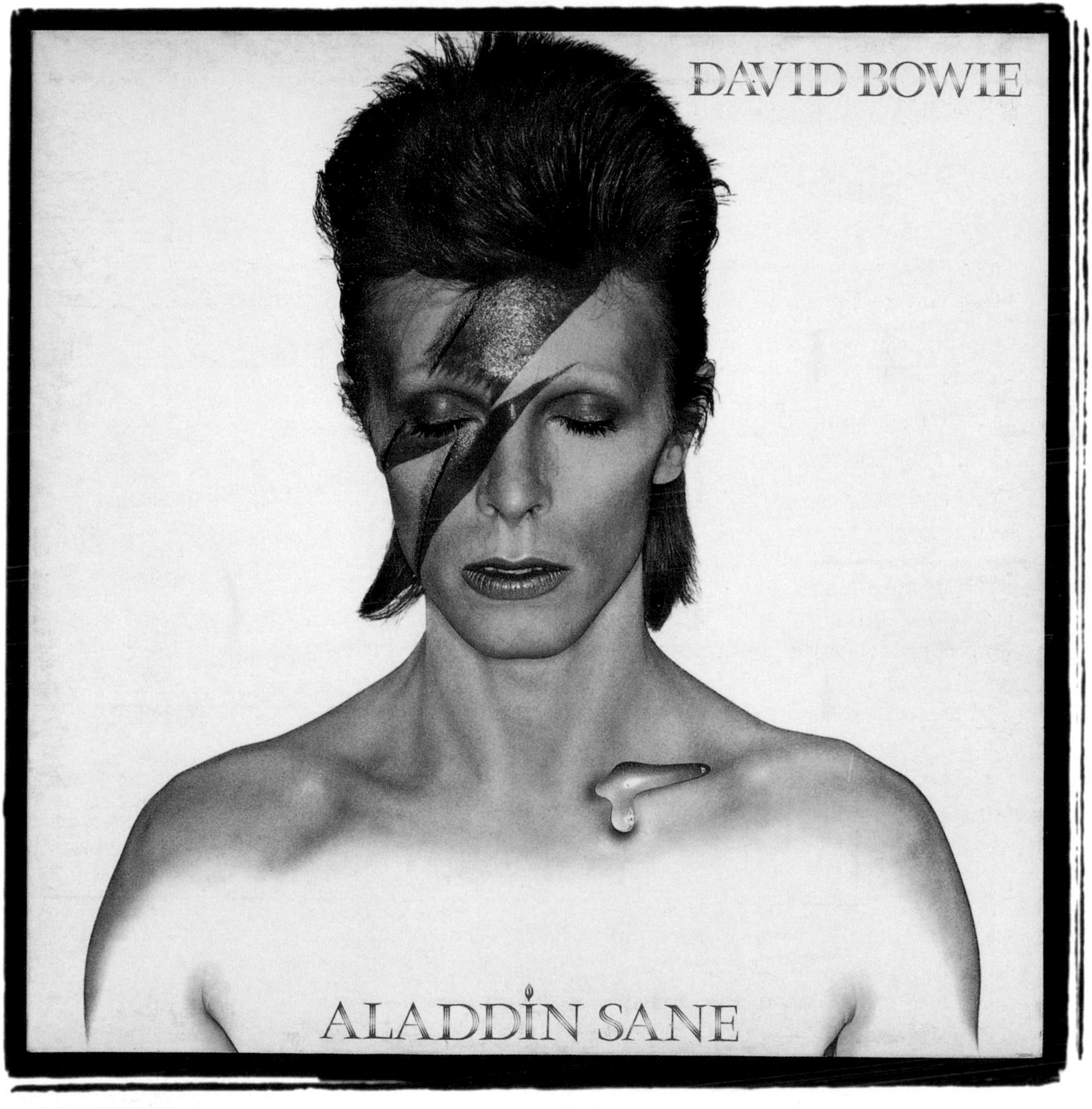
DAVID BOWIE
ALADDIN SANE

FUNKADELIC

COSMIC SLOP (1973)

Not until P-Funk mainman George Clinton hooked up with artist Pedro Bell did Funkadelic's image fully crystallize. They were *weird* before (see **Maggot Brain**'s [1971] sleevenotes, written by the Process Church of the Final Judgment), but Bell's artwork truly sent them into the stratosphere. With one foot in the Robert Crumb camp (the inner sleeve designs included a drawing for each song), Bell's artwork played on the P-Funk image of being some sort of futuristic sex-mongers coming to earth to tell people: 'Free your mind and your ass will follow.' The gatefold sleeve featured space aliens, dials for nipples and a booty ripe for shaking. Most importantly, though, it marked Funkadelic out as the biggest freaks around, an image Clinton was keen to cultivate, while making sure it was all taken in good humour with its comic-book design. Bell's outlandish – and some considered sexist – art would later cause trouble with **The Electric Spanking Of War Babies** (1981), which took the sci-fi sex theme further by having a woman lay naked on a phallic table while being spanked by paddles. For **Cosmic Slop**, though, Bell would modestly recall, '[It] wasn't my creative best, but it was my first one – so I think that was a nice start.'

RECORD LABELS

Westbound

RELEASE DATES

July 1973

SONGWRITERS

Sidney Barnes, George Clinton, William Franklin, Eddie Hazel, Cordell Mosson, Gary Shider, Bernie Worrell

NEW YORK DOLLS

NEW YORK DOLLS (1973)

So shocking was the New York Dolls' debut in 1973 that it was banned in Spain upon its original release. Glorification of street life and hard-drug culture can be almost entirely traced back to early 70s New York, with Lou Reed's **Transformer** (1973) being as key an influence as the New York Dolls' debut. The city was a melting pot for punk, hip hop and early dance music, and the New York Dolls brought a different kind of rock attitude overground, with a stark androgyny much more dangerous than anything 'glam' in the UK at that time. Derided by many, but changing the lives of the likes of Morrissey, NYD exuded an NYC cool that many still aspire to today. Their thin, ambiguous look, all made up with plenty of places to go, was at turns threatening and liberating for young kids just picking up guitars as punk slowly came to fruition. Like the Velvet Underground, the New York Dolls never achieved great commercial success at the time, but their look would influence practically every late-70s/early 80s new-wave group, and most of the 80s hair metallers that came after.

RECORD LABELS

Mercury

RELEASE DATES

August 1973 (US: July 1973)

SONGWRITERS

Principal songwriters: David Johansen, Arthur Kane, Sylvain Sylvain, Johnny Thunders

Secondary songwriter: Bo Diddley

STEVE MILLER BAND

THE JOKER (1973)

If you take a look at Steve Miller's sleeves from 1968 to 1972, they are all po-faced muso affairs. Blues rocker Steve was obviously fixing to be taken seriously in the glut of blues-boom musicians from the 60s, and with such sleeves he seemed to be pushing away anyone who was coming for good times and perhaps even a pop hook. That all changed with **The Joker**, however, which sees Miller step out with a Zal Cleminson-style mask, leather jacket and camp pose, almost as a 'Here I am, take me now!' to the fickle pop market. It is interesting that he wears a mask, though. **The Joker** was not Miller's artistic highpoint, but the title track marked his commercial peak when released as a single. It is almost as if Miller yearned so much for commercial success after so long that he felt he ought to take it, but hid behind the joker's mask so as to preserve some dignity. And so the follow-up, 1976's **Fly Like An Eagle**, returned to blues-rock design.

RECORD LABELS

Capitol

RELEASE DATES

October 1973

SONGWRITERS

Principal songwriter: Steve Miller

Secondary songwriters: Chuck Calhoun/Charles E. Calhoun, Eddie Curtis/Ahmet Ertegun, Obie Jessie/Sam Ling, Woody Payne

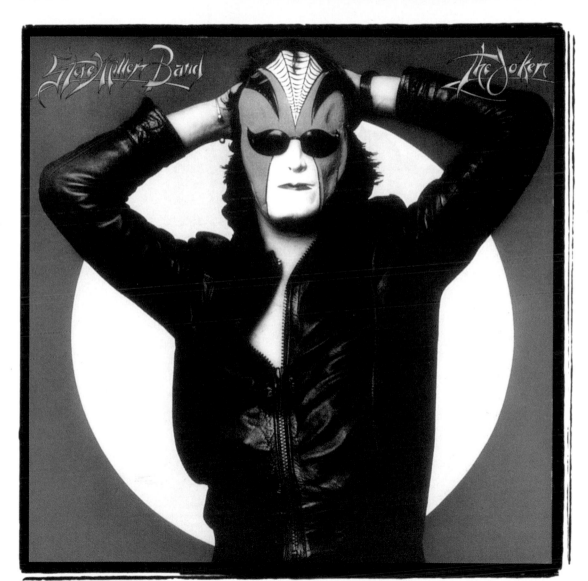

ELTON JOHN

GOODBYE YELLOW BRICK ROAD (1973)

Elton John's genius in the early 70s was his ability, along with collaborator Bernie Taupin, to write songs that spoke to the emotions and sensibilities of each and every one of his listeners. Simple imagery coupled with a wonderful knack for melody made his tunes catchy and accessible, while he also had a great handle on his image (though some may disagree). Previous album sleeves had seen John reference West-End musicals, but **Goodbye Yellow Brick Road** was his most ambitious to date. Even the poster that John is seemingly stepping into seems to be covering up an advertisement for a musical, but here John is inviting us to follow him (or his music) in order to escape from dull, repetitive factory life. While his image would get gaudier (and more overwrought) in due course, here he perfectly pays homage to music hall, show tunes, glam rock and one of the most identifiable movies of all time (thus, again, tugging at his listeners' emotions), in a painting that is as postmodern as it is nostalgic, creating something new out of all the old things it looks back on.

RECORD LABELS

DJM, MCA

RELEASE DATES

October 1973

SONGWRITERS

Elton John, Bernie Taupin

EMERSON, LAKE & PALMER

BRAIN SALAD SURGERY (1973)

For a band whose 70s catalogue is littered with fairly ropey album sleeves, **Brain Salad Surgery** jumps out as something so ornate and carefully done as to make you wonder why Emerson, Lake & Palmer didn't make a similar amount of effort for all of their sleeves. It is fitting that ELP's outstanding sleeve came with what is often regarded as their outstanding album. The working title for the album was actually *Slap Some Skull On Me*, which explains the artwork's complete lack of relevance to so-called brain salad surgery. The rather frightening machine contraption quite literally does 'slap some skull' down on the face of a sleeping woman, who is revealed when opening the gatefold sleeve. The window cut out from the centre of the front sleeve, giving a peek at the woman's lips beneath, adds to the unnerving concept. It is not surprising that sleeve designer H.R. Giger later designed the aliens for the movie *Alien* (1979); this incredibly detailed drawing is perfectly in keeping with such meticulous sci-fi design, and the album itself has been described as 'sci-fi rock'.

RECORD LABELS

Manticore

RELEASE DATES

December 1973

SONGWRITERS

Principal songwriters: Keith Emerson, Greg Lake

Secondary songwriters: William Blake/Charles Hubert Hastings Parry, Albert Ginastera, Peter Sinfield

DAVID BOWIE

DIAMOND DOGS (1974)

Quite literally the dog's bollocks … or not, as RCA had Bowie's half-human-half-hound genitalia airbrushed out of the original **Diamond Dogs** artwork, making the rare few that escaped castration much sought-after by collectors these days. Glam rock was on its last legs in 1974 and **Diamond Dogs** is something of a farewell to the scene. The decay is evident in the music, partially salvaged from an aborted musical based on George Orwell's *1984*, while the sleeve is much more gutter-glam than anything bright, spangly or celebratory. The distinctive art style came courtesy of Guy Peellaert, who found fame in 1972 with his artwork for the book *Rock Dreams*, in which he created made-up scenarios, such as the Rolling Stones dressed as Nazi Stormtroopers surrounded by young girls. Allegedly, Mick Jagger introduced Bowie to the book after commissioning Peellaert to design the sleeve for the Stones' forthcoming LP **It's Only Rock'n'Roll**. Bowie beat them to it, releasing **Diamond Dogs** in April 1974, making the Stones look like they copied him when their LP was released in October the same year.

RECORD LABELS

RCA

RELEASE DATES

May 1974

SONGWRITERS

David Bowie

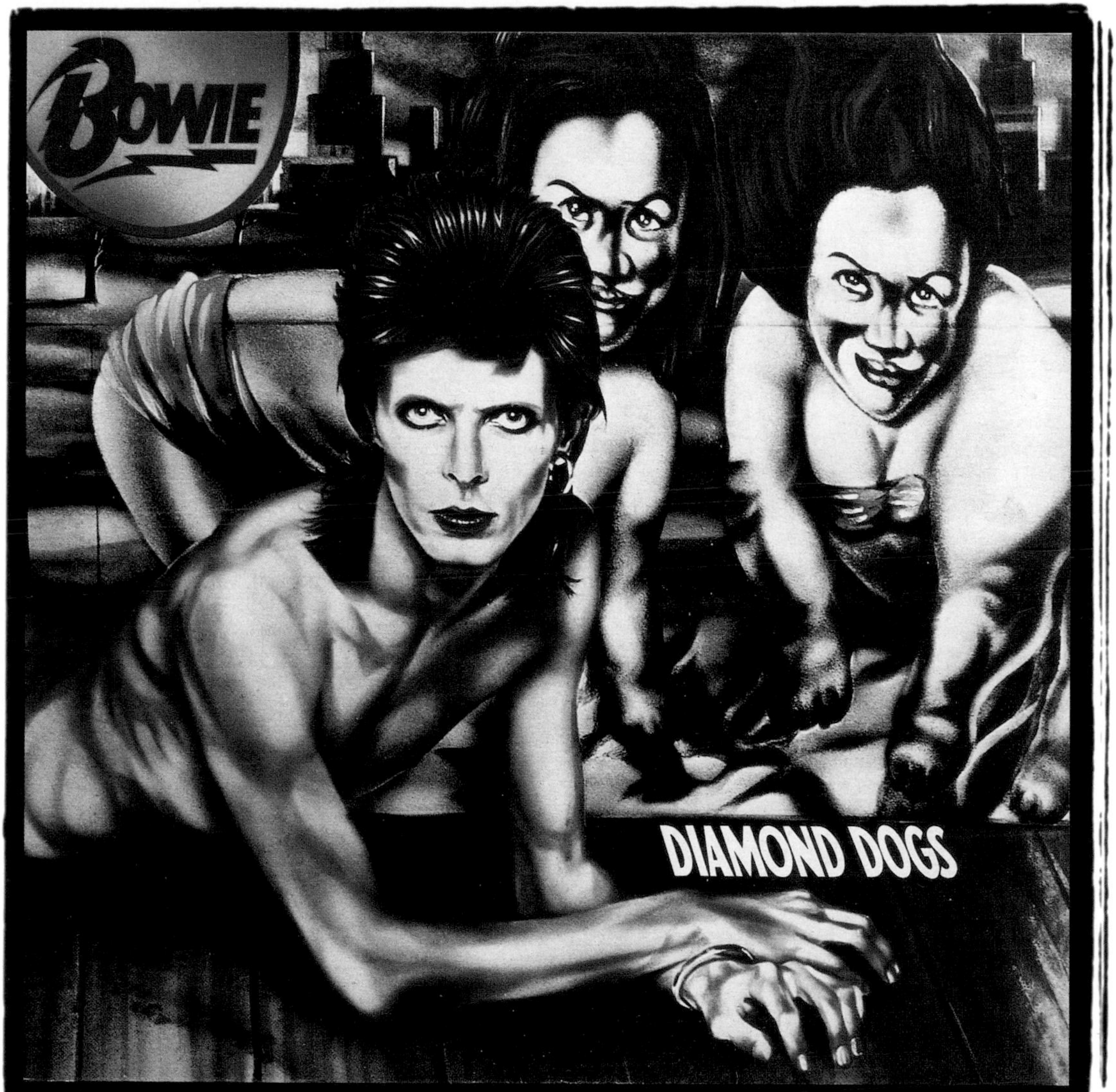
BOWIE
DIAMOND DOGS

LED ZEPPELIN

PHYSICAL GRAFFITI (1975)

By now masters of the creative album sleeve, Led Zeppelin were able to combine the gimmicky artwork that accompanied their album **Led Zeppelin III** (1970) (a sleeve with a hole cut out of it and a wheel with different images on it so that, when turned, it would reveal a different object in the hole at any one time) and the striking features of their best sleeves, such as **Houses Of The Holy** (see page 125) or their fourth, 'untitled', album (1971). For this double album the group came up with an outer-sleeve photo of buildings on St Mark's Place in New York City (Mick Jagger is also seen in front of them on the Stones' 'Waiting On A Friend' video). Each window of each building was cut out and the inner sleeves that contained the records had a number of different images printed on them so that, when placed inside the outer sleeves, they would appear in the buildings' windows; these included shots of the band dressed in drag. It had been almost two years since **Houses Of The Holy** was released and perhaps Jimmy Page, who baulked at …**III**'s gimmicky sleeve at the time, was more open to the idea of attracting attention any way he could with this sleeve.

RECORD LABELS

Swan Song

RELEASE DATES

March 1975

SONGWRITERS

Principal songwriters: John Bonham, John Paul Jones, Jimmy Page, Robert Plant

Secondary songwriter: Ian Stewart

CARLY SIMON

PLAYING POSSUM (1975)

Many will attest that **Playing Possum**, Carly Simon's fifth album, was actually something of a creative disappointment. Though Simon had seemed well aware of her physical attraction on sleeves in the past, perhaps she was falling back on that failsafe 'sex sells' marketing ploy this time round, to ensure that the album didn't go unnoticed. What is important to remember is that, though there had been the 'sexual revolution' of the 60s, the media didn't really catch up with that until the likes of Prince and Madonna leapt into the spotlight in the 80s. The image of Simon on her knees in lingerie and leather boots was not something people were expecting – especially from a mother. It seems that alcohol is to blame, together with photographer Norman Seef, as Simon took her normal daytime wear off and danced uninhibited for the camera after a few glasses of wine. The music might not have pushed the envelope, but in this case the sleeve itself was something of a minor revolution.

RECORD LABELS

Elektra

RELEASE DATES

May 1975

SONGWRITERS

Principal songwriter: Carly Simon

Secondary songwriters: Jackob Brackman, Billy Mernit, Mac Rebennack/Alvin Robinson

PiNK FLOYD

WiSH YOU WERE HERE (1975)

By 1975 Pink Floyd had learnt to deal with absence. With **The Wall** (1979) fast approaching and Roger Waters already feeling beaten down by the madness of the touring lifestyle after becoming an idolized rock star, it is not too much of a stretch to imagine him wondering where the real Roger Waters began, and whether he had become lost within the rock persona. For **Wish You Were Here** the group once again teamed with Hipgnosis to create a sleeve based around the theme of absence, and initially wanted the record to be packaged in entirely black wrap, as if it didn't even exist on the record-store shelves. Their record company baulked, however, and insisted that some sort of identifiable sticker be present, so that buyers would know the album was Floyd's. The flaming businessman, shaking hands with another businessman, hints again at vacancy. In this case, the businessman on the left should be offering help, not false business greetings and smiles. Spookily, Floyd's most obvious absent member, Syd Barrett, turned up at the studio during the recording of the remaining band members' tribute to him, 'Shine On You Crazy Diamond'. It was the last time he would be in the recording studio with his former band.

RECORD LABELS

Harvest, Columbia

RELEASE DATES

September 1975

SONGWRITERS

David Gilmour, Nick Mason, Roger Waters, Richard Wright

BRUCE SPRINGSTEEN

BORN TO RUN (1975)

The voice of several generations has created some of the most instantly recognizable sleeves with **Born To Run** and 1984's **Born In The USA** (see page 259). Of all the poses planned for the album's photoshoot, the image that was finally chosen for the sleeve was a pose that wasn't even meant for a photo. Springsteen just lent on E Street Band saxophonist Clarence Clemons for a second, but photographer Eric Meola happened to capture it and recalled, 'That one just sort of popped. Instantly, we knew that was the shot'. Springsteen's blue-collar rock may not mark him out as a man with a deft touch in many minds, but the thin lettering of the text was something of a design masterstroke at the time, giving the sleeve a chic quality that juxtaposes with Springsteen's rock roots. A version with a more scripted text exists and, amazingly, commands prices of up to £1,000. As with rock's most iconic images, **Born To Run** has been parodied countless times, notably by two children's TV heroes: *Sesame Street* and a parody within a parody of Fred Flintstone-based character Bruce Springstone singing '(Meet The) Flintstones'.

RECORD LABELS

CBS, Columbia

RELEASE DATES

October 1975 (US: September 1975)

SONGWRITERS

Bruce Springsteen

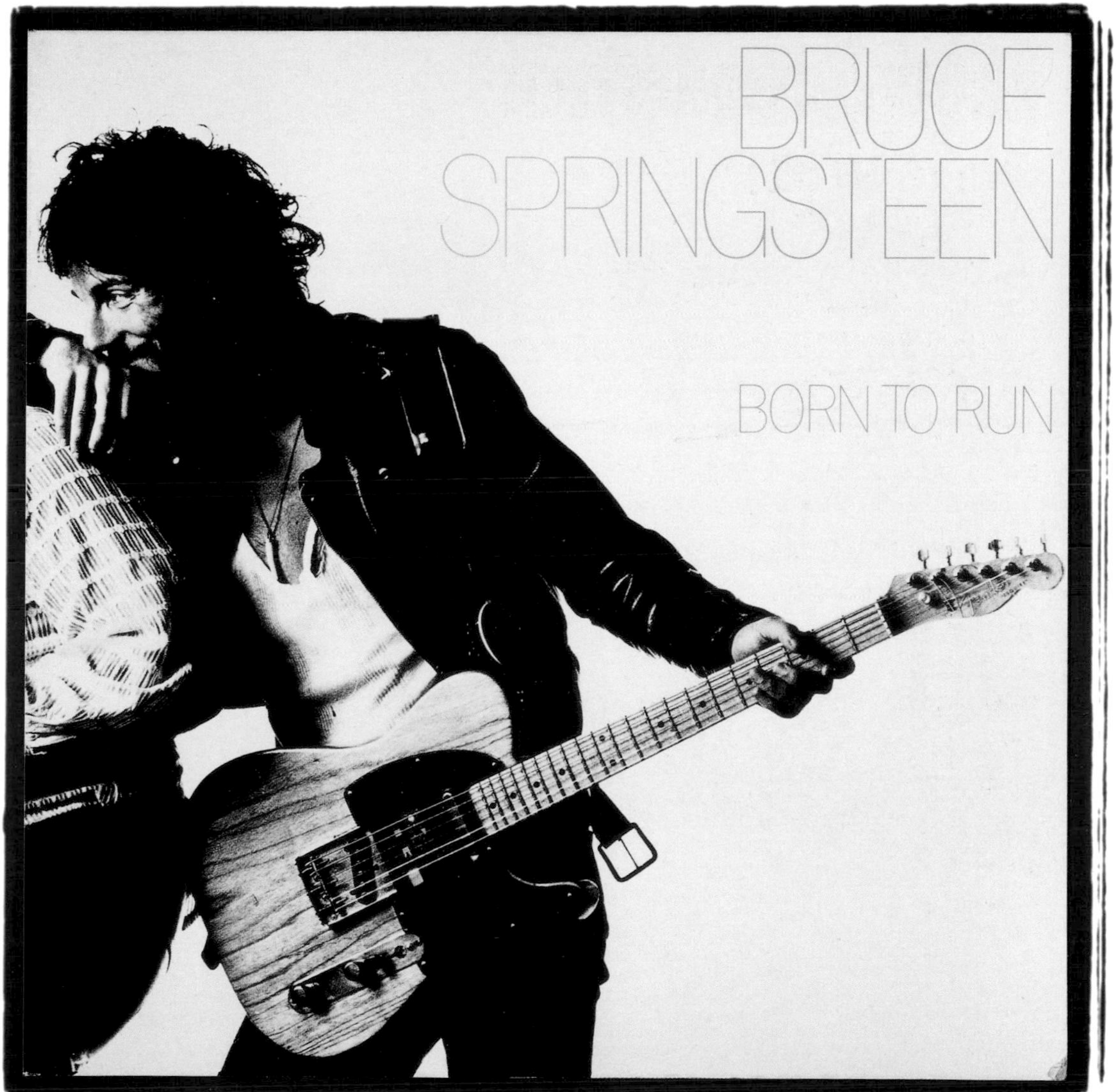
BRUCE
SPRINGSTEEN
BORN TO RUN

ROXY MUSIC

SIREN (1975)

Sirens were part of Greek mythology, sea demi-gods that sat on an island of rocks and sang out to sailors who, unable to resist their song and charms, would drive their ships towards the sirens in a frenzy, ultimately getting shipwrecked on the island. And so we come to Jerry Hall, a bewitching lady whom no man could resist, not even singers themselves. At the time of this, Roxy Music's fifth album, Hall was Bryan Ferry's girlfriend, so it stood to reason that he would have her feature in the line of iconic Roxy album sleeves (she was also in his 'Let's Stick Together' promo video the following year). It is probably no coincidence, either, that this was the album that opened with 'Love Is The Drug'. Sadly, the relationship wouldn't last. While Ferry was out touring with Roxy Music, Mick Jagger also found himself unable to resist Jerry Hall's charms and was wooing her with flowers and penning songs such as 'Miss You' for her. The bewitching Jerry Hall was later a spokesperson for Levitra, an anti-impotence medicine.

RECORD LABELS

Island, Atco

RELEASE DATES

October 1975

SONGWRITERS

Bryan Ferry, Eddie Jobson, Andy Mackay, Phil Manzanera

ROXY MUSIC

PATTI SMITH

HORSES (1975)

Most artists who search for a more spiritual plane than the one fame can offer – and Patti Smith is most certainly one of those – are not all that comfortable in front of the camera. Photoshoots need a level of willing posing and self-awareness that the likes of Bob Dylan would grow increasingly tired with as time went on, but which groomed and pampered pop stars are pretty much trained to do at will. Luckily for Patti Smith, she had been living with photographer Robert Mapplethorpe for five years before this shot was taken, resulting in an image that only the most intimate pairing could achieve. It is a shot that perfectly reflects the from-one-soul-to-another, stripped-down intensity of Smith's epochal rock'n'roll rebuttal. The stark black-and-white photo was as grainy and punk as they came in 1975, but Smith looks alternately delicately feminine and strongly masculine, her feminine figure and features contrasting with her masculine dress. Far from making herself look like a man to fit in with a man's world, here Smith is making male imagery fit her, and creating an iconic sleeve that looks like it has been tossed out with nary a thought.

RECORD LABELS

Arista

RELEASE DATES

December 1975

SONGWRITERS

Principal songwriters: Lenny Kaye, Ivan Kral, Patti Smith, Richard Sohl

Secondary songwriters: Chris Kenner, Allan Lanier, Van Morrison, Pete Townshend, Tom Verlaine

Patti Smith Horses

ARISTA

RAMONES

RAMONES (1976)

Essentially credited with kick-starting punk rock in 1976, the Ramones' first album delivered it all: short, sharp punk songs that lasted little more than two minutes, a whole album that comes in at under 30 minutes long, and a look that defined punk. It wasn't just the ripped jeans and leather jackets, either. The stone wall the group leaned against announced that this music came from street hoodlums, not cozy, privileged musicians with a limitless budget to spend and luxury views of the beach. In stark black-and-white, it stood out like Elvis Presley's debut (see page 21), defining itself by what it wasn't: it wasn't the work of a sensitive singer-songwriter, and it wasn't the overly elaborate, ornate work of a prog-rock band working with multi-track studio equipment. More than anything, this brought back the idea of the band as a gang. Though the members weren't brothers, they looked almost identical on the sleeve, leaning into each other and stressing togetherness. It wouldn't continue to be all roses for the group, but before punk became more of a fashion than a feeling, the Ramones embodied it in every way.

RECORD LABELS

Sire

RELEASE DATES

July 1976 (US: May 1976)

SONGWRITERS

Principal songwriters: Dee Dee Ramone, Joey Ramone, Johnny Ramone, Tommy Ramone

Secondary songwriter: Jim Lee

JONI MITCHELL
HEJIRA (1976)

Compare this sleeve to the twee, psych-infused one that accompanied Mitchell's 1968 debut **Song To A Seagull**, and we can tell that we have a very different Joni on our hands. Mitchell had been pushing herself further and further, even to the point of periodically tuning her guitar to a different key so that she had to re-learn all her old material, thereby keeping it fresh. By 1976 she had openly embraced jazz music, which helped her to stand out from the glut of mid-70s singer-songwriters. **Hejira** was a more polished and accessible album than her previous one, **The Hissing Of Summer Lawns** (1975), and the sleeve reflects that: no oblique imagery, but a large black-and-white photo of Joni, looking the epitome of jazz cool with her beret and cigarette. Much of the album deals with travel (she was on Bob Dylan's tour bus in 1975 when working up some of these tunes), and the freeway superimposed onto Mitchell's body seems to centre in on her core. As Mitchell herself has said, 'I wrote the album traveling cross-country by myself and there is this restless feeling throughout it…. The sweet loneliness of solitary travel.' Mitchell's sleeve gives hitchhiking an upper-class air removed from its 'wandering hobo' stereotype.

RECORD LABELS

Asylum

RELEASE DATES

November 1976

SONGWRITERS

Joni Mitchell

JONI MITCHELL HEJIRA

TOM WAITS

SMALL CHANGE (1976)

The masterpiece of Tom Waits' early phase (the down-on-his-luck, everyman barroom singer-songwriter era), **Small Change** embodied everything that Waits was mining in the mid-70s, stylistically and musically. The album was full of songs for hard drinking and broken hearts and, at the time, Waits was dangerously close to becoming the dirty, suit-wearing alcoholic so often portrayed in his songs. The sleeve presented Waits as the Bowery bum answer to Dylan's hepcat cool on **Bringing It All Back Home** (see page 41). As opposed to being surrounded by album sleeves signposting his musical past, though, Waits is trapped in a stripper's dressing room littered with reminders of the night before. Instead of looking out at the buyer with confrontation in their eyes, here the two look away, as if caught at an awkward moment. Who knows, maybe he has crossed the line and fallen in love with her; or perhaps he misunderstood her genuine affections and offered to pay. Waits lives in a complex world of human relations. While many singer-songwriters at the time were still clinging on to the pastoral imagery that supported their often airy conceits, Waits' sleeve looks more befitting of a 1950s pulp novel, and invites you to look into the seedier side of life.

RECORD LABELS

Asylum, Elektra

RELEASE DATES

May 1977 (US: November 1976)

SONGWRITERS

Tom Waits

Tom Waits   Small Change

THE EAGLES

HOTEL CALIFORNIA (1976)

By the mid-70s the true country-rock spirit of the late 60s and early 70s had well and truly given way to the more commercial offerings of those who merely adopted the image and sound. The Eagles, whose history as a collection of session musicians sees them continually donning the 'Hollywood cowboy' look (especially for their fantasy/myth-creating 1975 **Desperado** album sleeve), perfectly encapsulated this time for country music when they created one of its most successful, but arguably soulless, albums in **Hotel California**. The sleeve says it all: there is a decadence given to the Beverley Hills Hotel thanks to the lighting and the way the sun is setting behind. It is almost exotic, promising grand things of the place where they are 'livin' it up', but which could also 'be Heaven or … could be Hell'. As decadent as the sleeve – and title song – seems, there is a darkness and loss of innocence shining through in both, thanks largely again to the sun going down, perhaps on a lost youth. The album marked the peak of country rock's profitability (apart from the Eagles' first greatest hits, released earlier the same year), but also the moment when it breathed its last artistic breath.

RECORD LABELS

Asylum

RELEASE DATES

December 1976

SONGWRITERS

Principal songwriters: Don Felder, Glenn Frey, Don Henley, Randy Meisner, J.D. Souther, Joe Walsh

Secondary songwriters: Jim Ed Norman, Joe Viatle

FUNKADELIC

HARDCORE JOLLIES (1976)

Three years on from designing his first Funkadelic sleeve, Pedro Bell had got it down to a fine art. For Funkadelic's debut on Warner Bros. he afforded them his best – and possibly most commercially unfriendly – sleeve yet. The inside of the gatefold might have featured a cartoon pose of the band themselves, furthering their comic-book superhero look, but the outside continued the collision of graphic imagery and P-Funk ideals. Topless women aside, Funkadelic's past is referenced everywhere, not least in the character wearing an 'I Call My Baby Pussycat' top, after the song on 1972's **America Eats Its Young**. George Clinton was playing heavily on his past here: while there can be no mistaking what 'Hardcore Jollies' are, as it is an instrumental track, it is actually the live cut of 'Cosmic Slop' (originally recorded for the 1973 album of the same name, see page 129) that he has Pedro Bell focus on. In one manic drawing Bell illustrates the mother of the song, who turns to prostitution to feed her five children. The rest of the gatefold is, of course, pure P-Funk lunacy: steam trains, a man with a snake for a head playing guitar, and Batman and Robin running away from it all, scared.

RECORD LABELS

Warner Bros.

RELEASE DATES

February 1978 (US: 1976)

SONGWRITERS

George Clinton, Bootsy Collins, Grace Cook, Glen Goins, Michael Hampton, Gary Shider, Bernie Worrel

FUNKADELIC
(The U.S. FUNK MOB)
HARDCORE
JOLLIES
666
69
UP TO THE GOOD STUFF

PINK FLOYD
ANIMALS (1977)

Slightly continuing the theme of Pink Floyd's **Atom Heart Mother** sleeve (see page 95), **Animals** – for more obvious reasons – used another farmyard beast in its imagery. This time, a giant, helium-filled pig balloon was flown across London's Battersea Power Station, to create another iconic Storm Thorgerson-designed Pink Floyd sleeve. The photo shoot caused a lot of problems for the local authorities when the pig broke free of its moorings on the second day of shooting and began to fly over London. The flights due out of Heathrow Airport were delayed as the balloon got in the way of flight paths; the Civil Aviation Authority managed to track it until it reached 18,000 feet and went beyond their radar range. In the end it landed in a farmer's field and, after being repaired, was flown for a third time, enabling Thorgerson to get the shot of the pig over the power station. The image used for the sleeve itself was actually a composite of the pig as photographed on the third day of shooting, and the power station as captured on the first day.

RECORD LABELS

Harvest, Columbia

RELEASE DATES

January 1977 (US: February 1977)

SONGWRITERS

David Gilmour, Nick Mason, Roger Waters, Richard Wright

FLEETWOOD MAC

RUMOURS (1977)

By charting at No. 1, **Rumours** proved that Fleetwood Mac's second dawning as a commercially successful group wasn't just a one-off with the release of 1975's **Fleetwood Mac**. As if acknowledging that **Rumours** was worthy to walk in the footsteps of its predecessor, its sleeve is almost a 'Part Two' continuation of that album's artwork. Stevie Nicks, dressed up in her by then-*de rigeur* theatrical fairy costume, holds the crystal ball seen floating in the air on **Fleetwood Mac**, thrusting it out to the buyer, tempting them to purchase this follow up. Not only is this a strong visual trademark, but it also helps to cement Nicks's image as a somewhat mystical figure, possibly involved in witchcraft. Mick Fleetwood, of course, has some balls of his own here, serving to ridicule such nonsense. With marriages collapsing on all sides of the band at this point, rumours must also have flown about inter-band commitments and relationships, so Mick, the father figure of the group, was also no doubt saying 'bollocks' to that as well. It is no wonder that Nicks and Fleetwood were the only pair to pose for the sleeve; they were probably the only two band members happy to share a room.

RECORD LABELS

Warner Bros.

RELEASE DATES

February 1977

SONGWRITERS

Lindsay Buckingham, Mick Fleetwood, Christine McVie, John McVie, Stevie Nicks

IGGY POP

THE IDIOT (1977)

The most striking thing about Pop's first solo album is how far away it sets him from the self-destructive proto-punk madness of his early work with the Stooges. While the three Stooges LPs had sleeves that upheld the image of Pop as rock messiah, **The Idiot** removes him from punk in the year the music went overground, just as he could have been hailed as its figurehead. It was an inspired move that insured Pop's visibility, for the very reason that it wasn't expected. Longtime Iggy Pop fan David Bowie, who had done the original mix for the Stooges' **Raw Power** (1973), was essentially masterminding Pop's fledgling solo career, co-writing and producing **The Idiot**. It is no surprise, then, that as Pop was being moulded into something of a Renaissance man for the late 70s, somewhat in Bowie's image, his sleeve shot comes from the same source as that of **"Heroes"** (see page 173), which would be released seven months after **The Idiot**'s March 1977 release date. With this sleeve, the punk godfather turns himself into the post-punk godfather, along with the likes of Kraftwerk and Bowie, completely reversing his image in the process and wiping away the self-harming rock hedonist of yore … for a few years at least.

RECORD LABELS

RCA

RELEASE DATES

March 1977

SONGWRITERS

Principal songwriters: David Bowie, Iggy Pop

Secondary songwriter: Carlos Alomar

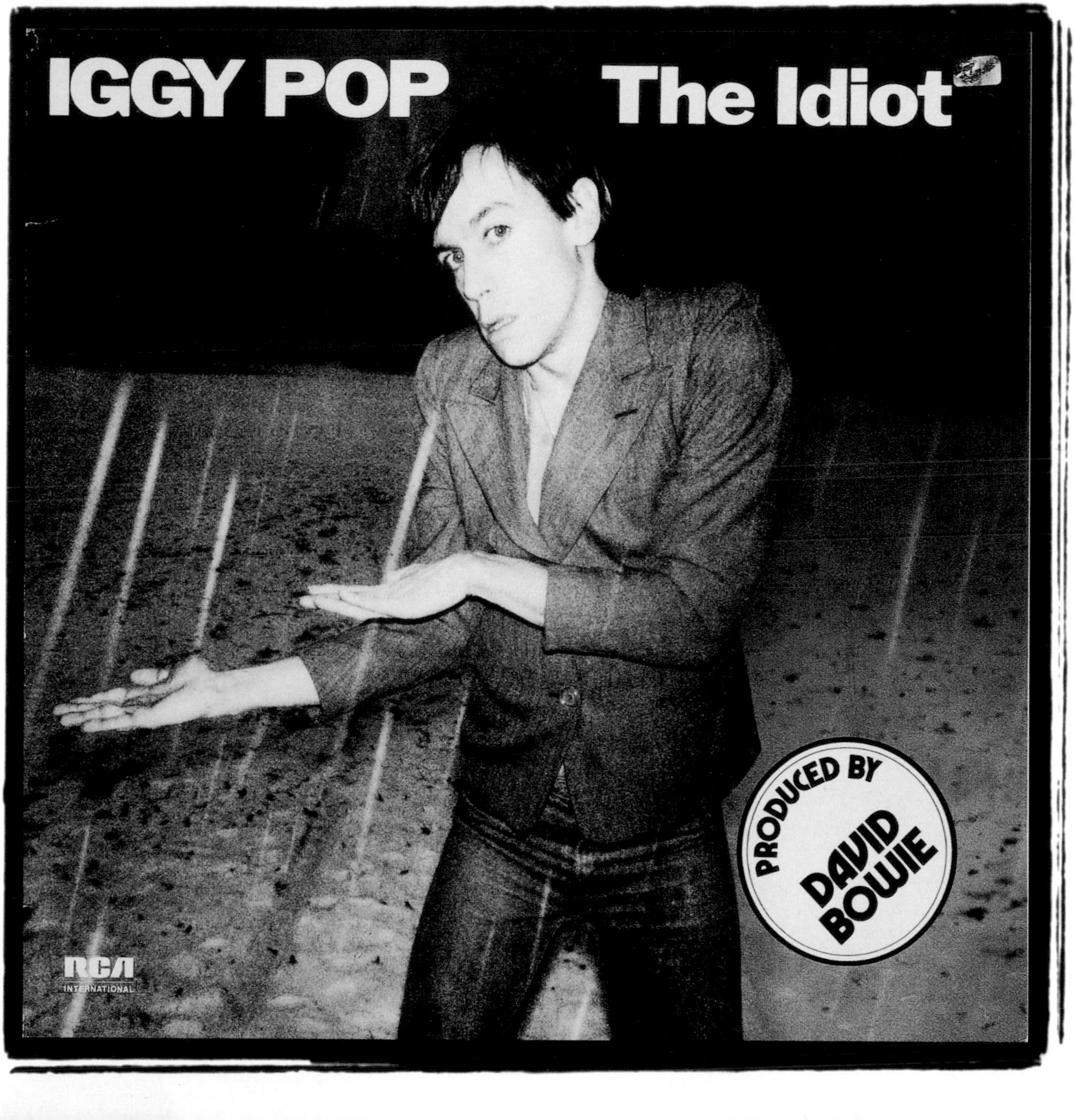
IGGY POP
The Idiot
PRODUCED BY
DAVID BOWIE
RCA
INTERNATIONAL

STEELY DAN

AJA (1977)

As Steely Dan's music was more complex, intelligently crafted and witty than that of many of their contemporaries', so their sleeve art set them aside as a band for whom intellect was usually prioritized over feel and spontaneity. **Aja** furthers Steely Dan's mystique, while essentially doing what it says on the sleeve: an Asian woman represents the album, the title of which is pronounced 'Asia'. But, as ever with Steely Dan, it is all about the nuances. Shot in a shadowed, disappearing profile, it forces the buyer to ask who this woman is (**Aja** was actually named after the Korean wife of Donald Fagen's friend's brother). Just as Steely Dan's music was textured, the simple-looking red and white of the woman's gown stands out against the black background in an almost three-dimensional way. In all, it is a remarkably direct sleeve, with none of the design flourishes of earlier Steely Dan album artworks, and it has a sleek sophistication wholly at odds with punk's gritty designs or prog's overblown works of art. As such, it is the perfect accompaniment to an album revelling in the band's composition skills, which also reached new heights of sophistication on this, their sixth LP.

RECORD LABELS

ABC

RELEASE DATES

September 1977

SONGWRITERS

Walter Becker, Donald Fagen

STEELY DAN

ABCL 268

IAN DURY

NEW BOOTS AND PANTIES!! (1977)

Although Ian Dury had been jobbing around the London music scene as early as 1974 with Kilburn & the High Roads, it wasn't until punk's year zero that he took to using his own name. Though certainly incorporating some punk elements into his sound and image (the black-and-white sleeve with brightly coloured text could be found on any number of punk sleeves at the time), Dury wasn't happy being a three-chord basher shouting along to polemic. Instead, he had an ear for melody and turned to succinct characterization in his songs, usually focusing on the more peculiar side of life, and most often settling on people's sexual habits: 'Billericay Dickie' deals with a womanizing man-about-town, while 'I'm Partial To Your Abracadabra' kind of speaks for itself. The sleeve sees Dury on a London high street, standing with his son Baxter, looking across the street to Woolworths (seen in the reflection of the shop window behind Dury). Dury, of course, isn't with the high-street chain at all, instead standing in front of rows of new boots and panties on sale in the kind of shop that his characters would probably frequent. That it also offers 'Large sizes available' adds to the not-quite-fitting-into-the-mould image.

RECORD LABELS

Stiff

RELEASE DATES

September 1977 (US: April 1978)

SONGWRITERS

Principal songwriters: Ian Dury, Steve Nugent

Secondary songwriter: Chaz Jankel

COME AND LOOK ROUND. STAGE + TV SPECIALITIES LARGE SIZES AVAILABLE Lots OF JACKETS TROUSERS, SHOES, RAINCOATS, ETC. ETC. SUITS !! 38-50 FROM £10 LEATHER SHEEPSKIN

SIZES UP TO 50

FITTING ROOMS AVAILABLE

IAN DURY NEW BOOTS AND PANTIES !!

DAVID BOWIE

"HEROES" (1977)

At the time of recording **"Heroes"**, the second album in a string of three known as the 'Berlin trilogy', Bowie was arguably as influenced by European styles as he ever would be in his career. The shot of Bowie on the sleeve was inspired by *Roquairol*, a painting by German artist Erich Heckel, and the cold, angular pose perfectly mirrors the Krautrock-inspired music of the album. Here Bowie was extending the experimentation of his **Low** (1977) compositions, and though the album peaked at No. 3 in the UK charts, the singles – 'Breaking Glass', 'Beauty And The Beast' and the title track – were some of his least successful in the UK since he had tasted superstardom. Perhaps the public just wasn't ready to accept a stark, black-and-white Bowie so at odds with the glam-rock hero he had become famous for being. Bowie was also producing Iggy Pop's solo work at the time of **"Heroes"**, and the sleeve of Pop's 1977 album **The Idiot** (see page 167) is another black-and-white take on *Roquairol*.

RECORD LABELS

RCA

RELEASE DATES

October 1977

SONGWRITERS

Principal songwriters: David Bowie, Brian Eno

Secondary songwriter: Carlos Alomar

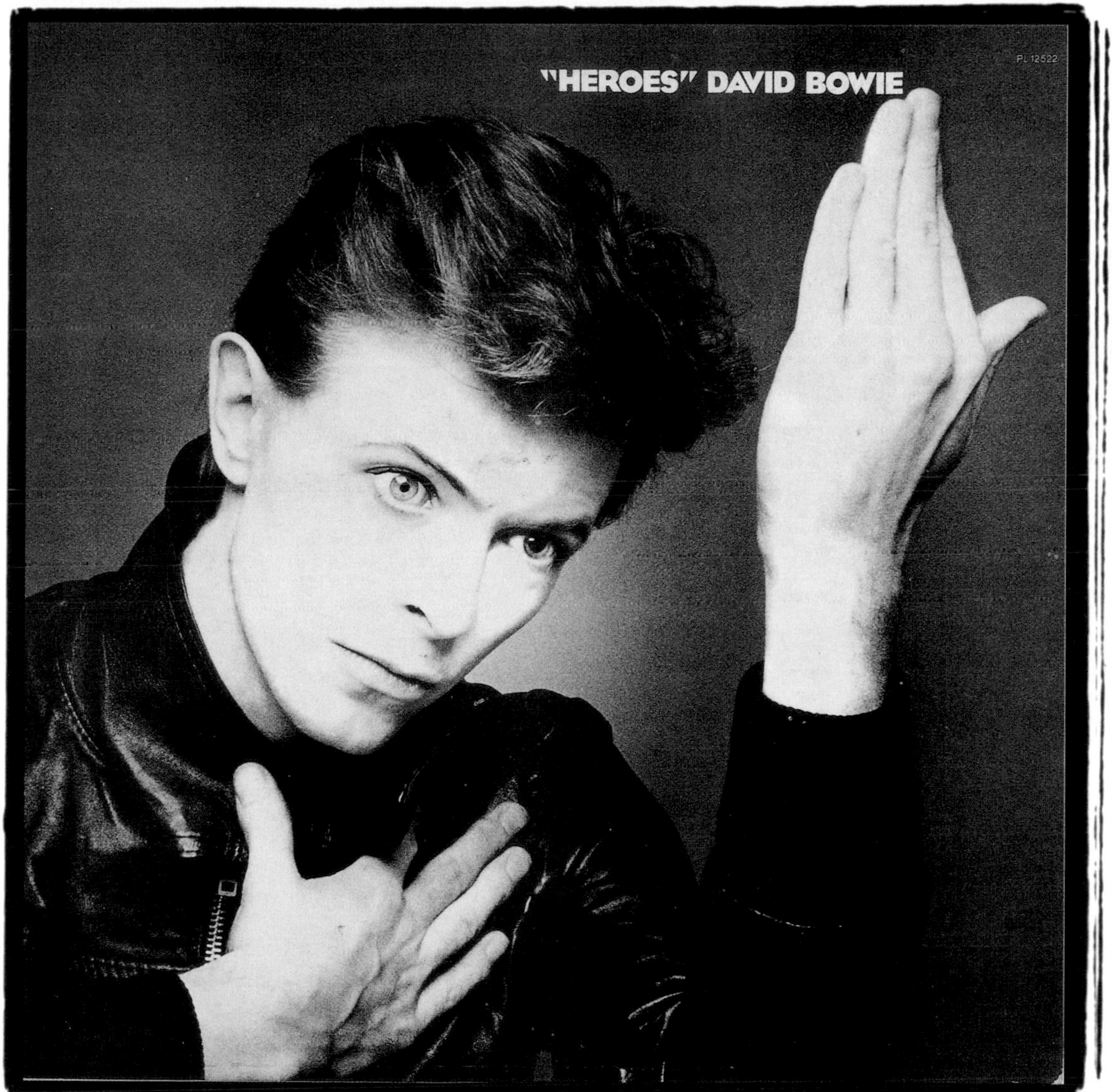
PL 12522
"HEROES" DAVID BOWIE

LYNYRD SKYNYRD

STREET SURVIVORS (1977)

Lynyrd Skynyrd's story is one of the most tragic in rock'n'roll history. Three days after the release of this, their fifth studio album and most successful so far, the band were in a horrific plane crash that killed lead singer Ronnie Van Zandt and guitarist Steve Gaines, while severely injuring the rest of the group. In an instant, one of America's most promising Southern-rock bands was brought to a crushing halt far too early. In a spooky coincidence, this original sleeve for **Street Survivors** features a photo of the band surrounded by flames. Even more unsettling is the fact that Steve Gaines is clearly standing out of line with the rest of the group, his head almost entirely engulfed by the fire. Though the original sleeve is back in use on modern CD releases of the album, the LP sleeve was quickly changed to feature the full band line-up starkly shot by a single spotlight, with a black background behind them; an image more befitting the tragedy that had befallen them.

RECORD LABELS

MCA

RELEASE DATES

October 1977

SONGWRITERS

Principal songwriters: Allen Collins, Steve Gaines, Gary Rossington, Ronnie Van Zandt

Secondary songwriter: Merle Haggard

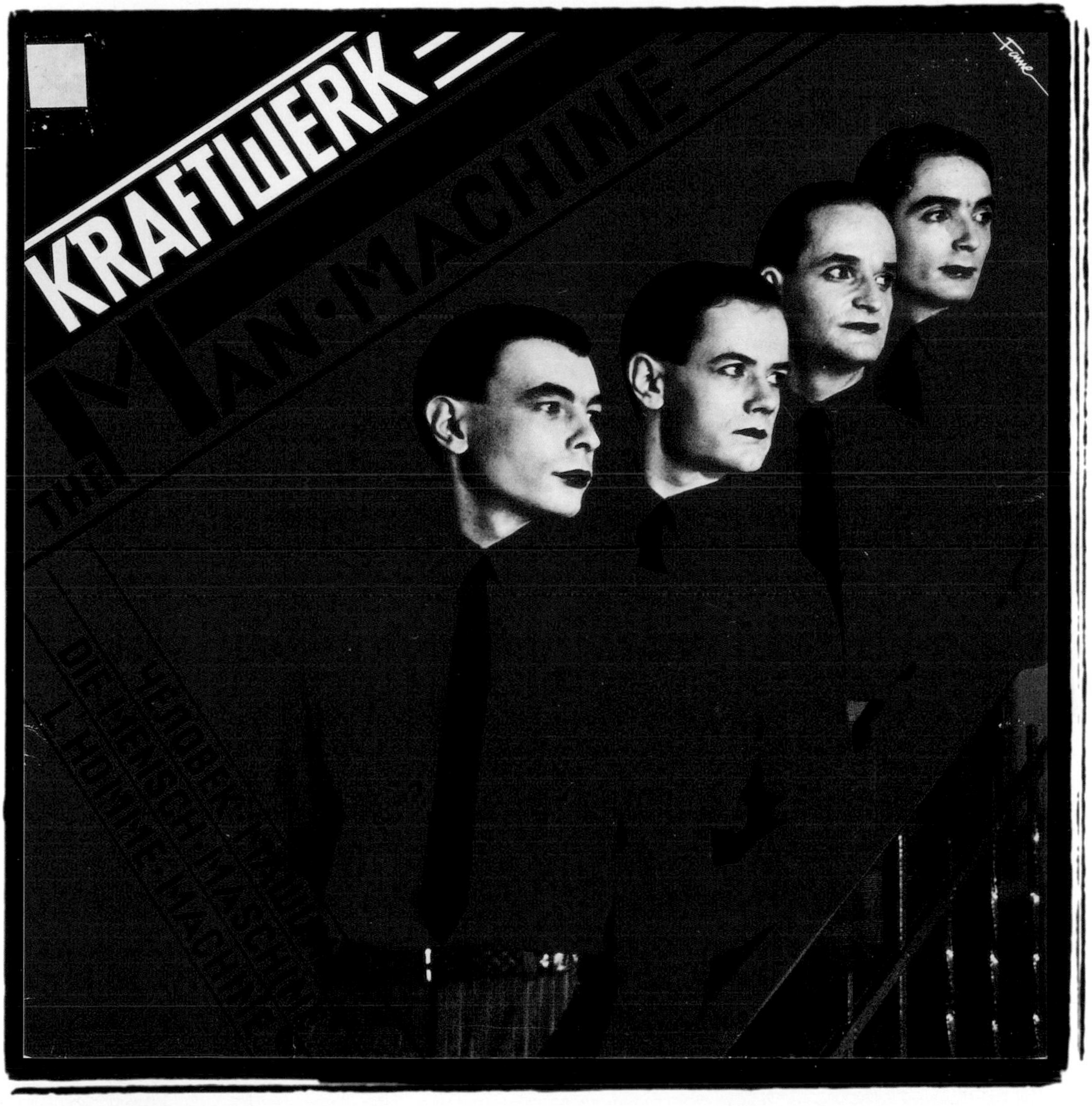
KRAFTWERK
THE MAN·MACHINE
Fame

GRAHAM PARKER & THE RUMOUR

THE PARKERILLA (1978)

There are many ways to say 'fuck you' to a record label. Graham Parker decided to release a rubbish live double album focusing on his only slightly better previous album, **Sick To Me** (1977), and in the process showed his record label, Mercury, that he really wasn't interested in them anymore (if there were any doubts after **The Parkerilla**, the follow-up single 'Mercury Poison' should have made it clear). As half-assed as the live album seemed to be, at least Parker turned up to play on it. The gatefold sleeve sees him effectively play dead on what looks like it could be a coroner's table, shot in black-and-white. It probably took about three seconds to come up with the idea, and a further three to take the photo and add the blood specks. Compare this with his follow-up, **Squeezing Out Sparks** (1979), on the Arista label, and the idea is writ large: from its title to its bright, more lively cover (Parker looks almost electrocuted), **Squeezing Out Sparks** sees Parker very much alive and ready to record. **The Parkerilla**, however, should remain dead and buried.

RECORD LABELS

Vertigo, Mercury

RELEASE DATES

May 1978

SONGWRITERS

Graham Parker

THE ROLLING STONES

SOME GIRLS (1978)

By 1978 the Stones were more than a decade-and-a-half old and faced the problem of competing with the late 70s' most popular music: punk/new wave and disco. Ron Wood joined the band full-time at this point, giving Keith Richards a much more dynamic sparring guitar partner than Woods' predecessor, Mick Taylor, had been, and providing the Stones with some new, punkish, energy. For his part, Mick Jagger had been enjoying the jet-set lifestyle and was courting the disco audience that had come over from New York. As such, the big single lent its title to the album, and the concept for the sleeve. A die-cut outer sleeve, not dissimilar to that of Led Zeppelin's **Physical Graffiti** (see page 141), was designed, with an inner sleeve featuring pictures of the Stones that could be placed behind the holes. Putting the two together, you could have the Stones' faces pop up in place of various female celebrities, though they would run into trouble when the likes of Raquel Welch and Marilyn Monroe (as represented by her estate) threatened legal action, causing their images to be removed on subsequent pressings.

RECORD LABELS

Rolling Stones

RELEASE DATES

June 1978

SONGWRITERS

Principal songwriters: Mick Jagger, Keith Richards

Secondary songwriters: Barrett Strong, Norman Whitfield

BLONDIE

PARALLEL LINES (1978)

Blondie's third album marked the band out as pure, slick pop stars, somewhat leaving behind the dying new-wave scene. Gone was the street imagery of predecessor **Plastic Letters** (1978), and in was a purely aesthetic sleeve, singling Debbie Harry out as the marketable face of the group. Though Blondie's songs were written by all band members, the boys' suits merge them into one entity (and into the striped background) with one foot in the post-punk look, while Harry steps out as the glamorous representative of Blondie. In truth, Harry had always taken to two-tone clothing, even in Blondie's earliest days, and this iconic image was the culmination of years of image-making on her part. The parallel lines speak for themselves, while the album was titled after a poem written by Harry, but which was never recorded as a song. It has also been argued that the parallel lines, as lines which never join up with each other, mirror the subject matter of most of the songs, which deal with human relations that don't tie up either. Oddly enough, Blur parodied the album sleeve in the mid-90s, with Damon Albarn stepping out as Debbie Harry.

RECORD LABELS

Chrysalis

RELEASE DATES

September 1978

SONGWRITERS

Principal songwriters: Jimmy Destri, Nigel Harrison, Debbie Harry, Frank Infante, Jack Lee, Chris Stein

Secondary songwriters: Joe B. Mauldin/Norman Petty/Niki Sullivan

JOE JACKSON
LOOK SHARP! (1979)

As the 70s started to turn into the 80s, the backlash against overt punk imagery had to begin somewhere. With echoes of Michael Jackson's later **Moonwalker** logo, **Look Sharp!** not only mirrors its title, but sets Joe Jackson apart from his new-wave contemporaries. Though he began with one foot in the post-punk camp, like Elvis Costello it wasn't long before Jackson moved away from that scene and carved a niche out for himself as an 80s singer-songwriter, unafraid to try myriad styles. Rather than another black-and-white sleeve featuring a suited new waver staring out at the buyer, **Look Sharp!** features some of the more imaginative artwork of the year, even if it wasn't entirely conceived as such. The close-up shot of Jackson's shoes was largely achieved by accident. Jackson has recalled, 'I think we just wanted a photo of my Denson's shoes, because I'd just bought them the day before. [Photographer] Brian Griffin took maybe one or two shots of them as part of an entire day spent doing more typical portrait shots.'

RECORD LABELS

A&M

RELEASE DATES

January 1979

SONGWRITERS

Joe Jackson

JOE JACKSON

LOOK SHARP!

SUPERTRAMP

BREAKFAST IN AMERICA (1979)

Whether ironic or not, **Breakfast In America** doesn't go very far towards dispelling the myths surrounding Americans and their eating habits. Actress Kate Murtagh, best known for her roles on American TV, parodies the Statue of Liberty, dressed as a waitress with a glass of orange juice instead of a torch, and a diner menu in place of where the statue holds a tablet marking the date of American Independence. America may be the land of the free, but here it is also the land of you are what you eat, as the famous New York City skyline is reconstructed using a cornflake box, cutlery, a large plate of breakfast and other food paraphernalia, including a mustard bottle and egg cartons. The whole scene was constructed by sleeve designer Mike Doud and spray-painted white to look like the NYC skyline. Doud has commented that, since the band were not interested in appearing on their sleeves themselves, the image represents 'New York seen through the eyes of someone that sees it not as a gateway to the east of America but as a gateway to Route 66. It was a West-Coast treatment of an East-Coast icon'.

RECORD LABELS

A&M

RELEASE DATES

March 1979

SONGWRITERS

Rick Davies, Roger Hodgson

THE CURE

THREE IMAGINARY BOYS (1979)

Apart from being inherently weird – even for a Cure sleeve – perhaps the most interesting thing about **Three Imaginary Boys** is what it spawned, and how it sticks out like a sore thumb in the Cure's catalogue. For their debut album, the Cure's record label, Fiction, chose not only the sleeve, but also which songs were going to be on the album, all without Robert Smith's input. As a result, Smith has demanded total artistic control over all Cure releases since. It is not hard to see why. A tall lamp, refrigerator and Hoover are hardly staple Cure images. Perhaps it would have been an acceptable sleeve for the 'Love Cats' single, but stand this up next to the sleeves for the likes of **Pornography** (1982) or **The Head On The Door** (1985) and, for the archetypal post-punk/goth band, it is an embarrassment. Of course, the Cure had hardly hit their stride; the music is decidedly poppy, with pop-punk tunes more angular than the long, abstracted pieces they would come up with later. Perhaps, then, it is best for all concerned that **Three Imaginary Boys** sticks out. Anyone looking for doom and gloom would be better served with releases further down the line.

RECORD LABELS

Fiction

RELEASE DATES

May 1979

SONGWRITERS

Principal songwriters: Michael Dempsey, Robert Smith, Lol Tolhurst

Secondary songwriter: Jimi Hendrix

IAN DURY & THE BLOCKHEADS

DO IT YOURSELF (1979)

These days it is normal for a band to release albums and singles on multiple formats, with different artwork for each, and different artwork for each part of the world that the album is released in (see the White Stripes' **Elephant**, page 357, which had six variations on the sleeve for the six major world territories). Back in 1978, however, this sort of thing was not exactly common practice, so when Ian Dury (this time with backing band the Blockheads credited) released the follow-up to **New Boots And Panties!!** (see page 171) it sent hardcore fans into a frenzy, as they were faced with the task of collecting *12* different sleeves. In a promotional masterstroke (and one which cemented Dury's reputation as a chronicler of all things British), each of the 12 sleeves featured a different Crown wallpaper design. While Crown are now probably better known for their paint, in the 70s they designed all manner of promotional paint brushes and badges etc. These days it is equally common for a band to get into sponsorship deals with companies, though Crown might well be one of the oddest, considering the rock'n'roll excesses out there to be explored.

RECORD LABELS

Stiff

RELEASE DATES

May 1979 (US: July 1979)

SONGWRITERS

Charley Charles, Ian Dury, Mickey Gallagher, Chaz Jankel, John Turnball, Norman Watt-Roy

DO IT YOURSELF

IAN DURY
& the
BLOCKHEADS

SIDE ONE	SIDE TWO
INBETWEENIES	THIS IS WHAT WE FIND
QUIET	UNEASY SUNNY DAY HOTSY TOTSY
DON'T ASK ME	MISCHIEF
SINK MY BOATS	DANCE OF THE SCREAMERS
WAITING FOR YOUR TAXI	LULLABY FOR FRANCES

CROWN P97806

Tommy
THE TALKING   TOOL BOX SAYS
IT'S ... FOR ALL THE   FAMILY TO ENJOY!

THE CARS

CANDY-O (1979)

Taking a cue from the Roxy Music school of sleeve design, the Cars' **Candy-O** is another adhering to the 'sex sells' principle. In hindsight, TV channels such as *Men And Motors* show how close to the mark the Cars were with this, the sleeve to their second album. The model on the sleeve was also called Candy (Moore), and was going out with the drummer when the group recorded the album (Ric Ocasek claims that the 'O' stands for 'Obnoxious'). Most amazingly, the group convinced artist Albert Vargas to come out of retirement to create the painting. Vargas was a famous erotica painter, best known for his 'Vargas Girls' work for *Playboy*, who gave up painting after his wife died in 1974. That the car is represented as just an outline speaks volumes in a way that the band may not have intended; having the painting of Candy Moore (based on a photo taken at a Ferrari dealer's in Beverly Hills) standing out in full colour against the car's outline seems to concede that the band and music are playing second fiddle. What more do you want? It's got a semi-naked woman on it!

RECORD LABELS

Elektra

RELEASE DATES

June 1979

SONGWRITERS

Ric Ocasek

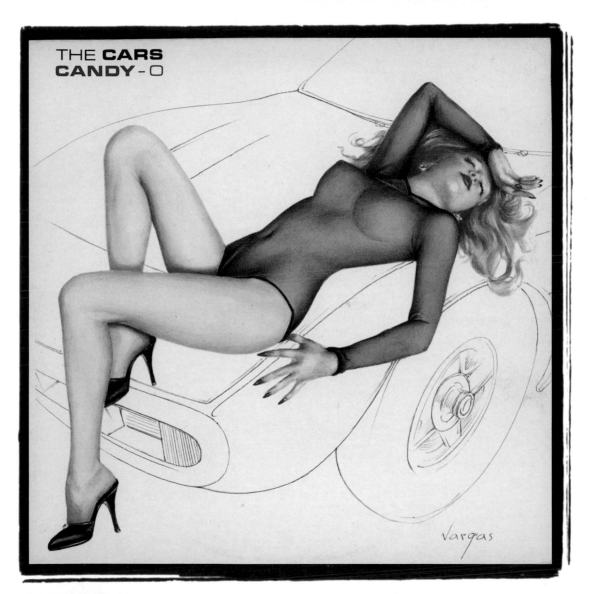

JOY DIVISION

UNKNOWN PLEASURES (1979)

Musically, lyrically and stylistically, Joy Division provided true new wave with its entire template. The group's somewhat glacial, synth-infused post-punk music continues to inspire swathes of bands, and their stripped-down sense of aesthetics is perfectly captured by **Unknown Pleasures**' sleeve. Artist Peter Saville is credited with designing the image, along with the band themselves and artist Chris Mathan. Taken from the *Cambridge Encyclopedia of Astronomy*, it is actually the radio wave of 100 pulses emitted by the first pulsar (a neutron star) found in space. Originally drawn as black lines on a white page, Saville et al reversed the colours and left the image at that. Adding to the band's mystique is the fact that neither the front nor rear sleeve had a band name or tracklisting, with just the title and catalogue number FACT 10 appearing on the back. Sadly, 2007 saw the production of 'Unknown Pleasures' trainers courtesy of New Balance: the sole featured the radio wave design, while 'FACT 10' was embroidered into the back of the trainer. And so another sacred cow gets slaughtered....

RECORD LABELS

Factory

RELEASE DATES

August 1979

SONGWRITERS

Ian Curtis, Peter Hook, Stephen Morris, Bernard Sumner

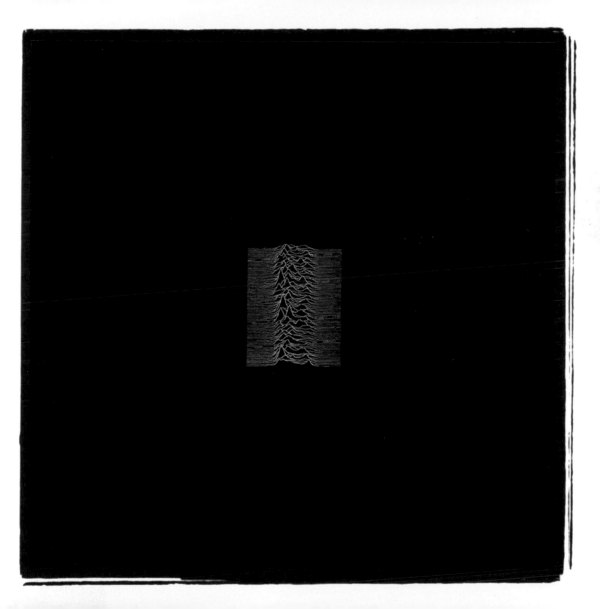

TALKING HEADS

FEAR OF MUSIC (1979)

Talking Heads were a band that were always looking to eradicate their image. Even when the band members did appear on their own album sleeves, it was often as distorted group shots that gave little clue as to what they actually looked like (see **Remain In Light**, page 223). For their third album, the Brian Eno-produced **Fear Of Music**, the sleeve was in keeping with much of the stark, new-wave imagery of PiL's **Metal Box** (see page 205) or Joy Division's **Unknown Pleasures** (see page 197): something which gave little concession to group image. Looking like the sort of metal flooring you would find in a factory, the original LP sleeve actually came embossed, so as to give the feel of the suggested material. The foreboding appearance is perfectly in keeping with the music. For followers of the band who knew Talking Heads' first two albums, they would have been shocked by **Fear Of Music**'s more 'difficult' listen. Byrne still clung to his quirky observations on human life, but stripped the music of much of its melody. **Fear Of Music** is almost saying 'enter if you dare' to those used to singing along to songs such as 'Psycho Killer'.

RECORD LABELS

Sire

RELEASE DATES

August 1979

SONGWRITERS

Principal songwriter: David Byrne

Secondary songwriters: Hugo Ball, Brian Eno, Chris Frantz, Jerry Harrison, Tina Weymouth

TALKING HEADS
FEAR OF MUSIC

MADNESS

ONE STEP BEYOND... (1979)

Make way for the Nutty Boys! Madness landed almost fully formed in 1979, with their 'cheeky cockney lads' image being pushed right from the start. Perhaps the least threatening-looking gang to come out of the post-punk era, they nevertheless presented themselves as a group of boisterous ragamuffins coming straight from the East London streets. Their ska pop instantly took hold, and it wasn't long before Suggs and co. would rub shoulders with the premier songwriters of the 80s. In the early days, however, their image almost overshadowed their true talents, and the 'nutty walk' featured here has defined them. Taking cues from the Monkees' walk, it is a simple enough gesture that instantly solidifies a gang mentality (although Chas Smash didn't feature on the front sleeve, as he was yet to become a full time band member). It also goes some way to capturing in image what went on in sound once you played the record, as the group gleefully sprang from pop to ska with no small degree of humour (see the likes of 'My Girl'). There are plenty of anodyne band shots in the world, but with **One Step Beyond...** Madness managed to bring humour back to the pop-group look in time for the coming decade.

RECORD LABELS

Stiff, Sire

RELEASE DATES

October 1979

SONGWRITERS

Mike Barson, Mark Bedford, Cecil Campbell, Chris Foreman, Shanne Hasler, Graham 'Suggs' McPherson, Carl 'Chas Smash' Smyth, Lee Thompson

STEVIE WONDER

STEVIE WONDER'S JOURNEY THROUGH THE SECRET LIFE OF PLANTS (1979)

Following Marvin Gaye's lead, Stevie Wonder had insisted upon having total artistic control over his work at Motown. Wonder's run of ever-inventive classics in the early 70s reaped rewards for both artist and label, starting with 1972's **Music Of My Mind** and ending with 1976's **Songs In The Key Of Life**. When he presented Motown with **…Journey Through The Secret Life Of Plants** three years later, however, after the longest wait yet for a new Stevie Wonder album, Motown were left scratching their heads. The pastoral sleeve is wholly at odds with the African roots-style artwork of the likes of **Innervisions** (1972) and **Talking Book** (1973), and those who looked to Wonder's music to reflect African-American life in the 70s were shocked by what appeared to be an album sleeve better suited for a cover-mount on a botany magazine. Wonder had never been one to remain pigeonholed, however, breaking with tradition was his intent. The mostly instrumental album was like nothing he had ever recorded, and formed the soundtrack to a documentary of the same name. In hindsight, it has been claimed that Wonder's album was actually the first new-age recording, but at the time the entire image was removed from everything the artist was seen to represent.

RECORD LABELS

Tamla-Motown, Tamla

RELEASE DATES

November 1979

SONGWRITERS

Principal songwriter: Stevie Wonder

Secondary songwriters: Michael Sembello, Syreeta Wright, Yvonne Wright

STEVIE WONDER'S

Journey Through

The Secret Life of Plants

PUBLIC IMAGE LTD

METAL BOX (1979)

Public Image Ltd afforded John Lydon the chance to reinvent himself after the Sex Pistols and lay the Johnny Rotten legend to rest for a while. In many circles, PiL's avant-garde post-punk dub noise rock is arguably more influential than anything Lydon did with his former band. With the artwork for their second album quite literally doing what it says on the tin, they also created one of the most innovative and memorable album packages since **Sgt. Pepper...** (see page 49), and one that perfectly chimes with post-punk industrialism. **Metal Box** came packaged in a round film canister-like metal tin, with just the 'PiL' logo to identify it. Amazingly, it didn't cost much more than a standard LP package to produce. Such packaging has proven popular in recent times, for example on a recent reissue of the Small Faces' **Ogdens' Nut Gone Flake** (see page 59). **Metal Box** was also notable for coming pressed on three 45 rpm 12-inch records, as opposed to just one 12-inch LP. Though more expensive than pressing 12-inches to run at a speed of 33 $\frac{1}{3}$ rpm (as LPs usually are), cutting them this way ensures better sound quality. For such a heady sonic mix as **Metal Box**, it was certainly a wise move.

RECORD LABELS

Virgin, Island

RELEASE DATES

December 1979

SONGWRITERS

Keith Levene, John Lydon, Jah Wobble

THE CLASH

LONDON CALLING (1979)

In parodying the sleeve for Elvis Presley's debut album (see page 21), the Clash were setting their stall out early on their third release. A mixture of punk, reggae and rockabilly, **London Calling** may well be to punk/post-punk music what **Elvis Presley** was to rock'n'roll – a collection of influences wider than the genres the artists were recognized as spearheading. Like the artwork it is based on, **London Calling** also perfectly captures the spirit of punk in the way that **Elvis Presley** captured rock'n'roll. Longtime Clash photographer Pennie Smith (so longtime that she photographed The Good, The Bad And The Queen, a supergroup of sorts featuring Clash bassist Paul Simonon alongside Damon Albarn, Tony Allen and Simon Tong) took the photo of Paul Simonon smashing his bass onstage at New York's Palladium during their 'Clash Take The Fifth' US tour in September 1979. Though it has gone down in history as one of the greatest rock'n'roll photos of all time, Smith originally didn't want to use the shot for the album's sleeve, as she felt it was too out-of-focus. Thankfully, Joe Strummer convinced her otherwise.

RECORD LABELS

CBS, Epic

RELEASE DATES

December 1979 (US: January 1980)

SONGWRITERS

Principal songwriters: Mick Jones, Joe Strummer

Secondary songwriters: Clive Alphonso, Topper Headon, Paul Simonon, Vince Taylor

206

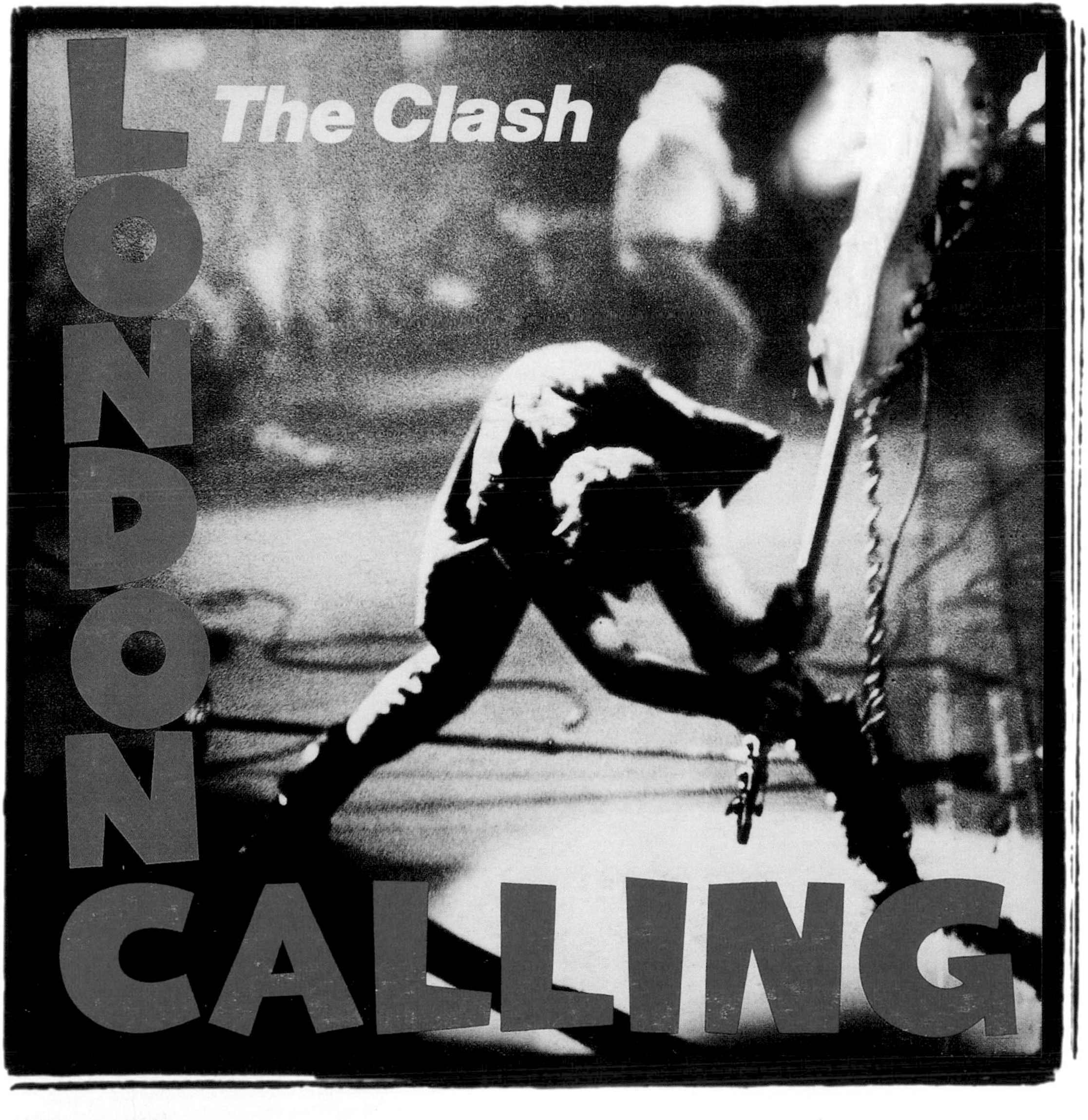
The Clash
LONDON
CALLING

ALBUM COVERS

RIFF
'ORY

EIGHTIES

209

BOZ SCAGGS

MIDDLE MAN (1980)

By the 1980s Boz Scaggs was a seasoned musician, having learned his trade playing with Steve Miller in the 60s and released a string of solo albums in the 70s. By **Middle Man**, his ninth solo album since 1965, Scaggs was an old hand at this sort of thing, and the album saw him straddling his singer-songwriter past, while embracing the new production techniques that the 80s had to offer. The sleeve saw him wholeheartedly step into the decade. A direct contrast to the rootsier singer-songwriter sleeves seen on the likes of 1972's **My Time**, **Middle Man** sees Scaggs pick up from where 1977's **Silk Degrees** sleeve left off, both with its clean-cut sleeve and more commercial rock appeal (Toto even co-wrote the album with him). Somewhat pointing the way towards Robert Palmer's 'Addicted To Love' image, **Middle Man** saw Scaggs presenting himself as the commercial face of rock music, much like the look Genesis would adopt during their **We Can't Dance** era. As make-up and flamboyant dress would signpost new romanticism and electro pop in the 80s, Scaggs' suited and booted look, with stockinged model in tow, pretty much defines 80s MOR (middle-of-the-road) rock. Even David Bowie would be wearing suits in three years' time.

RECORD LABELS

Columbia

RELEASE DATES

April 1980

SONGWRITERS

Davis Foster, David Lasley, Steve Lukather, Boz Scaggs, Bill Schnee

PETER GABRIEL

PETER GABRIEL (1980)

Often referred to as 'Melt', thanks to the distorted photo of Peter Gabriel on the sleeve (or 'III', because it was Gabriel's third solo release since leaving Genesis), the album artwork satisfied Gabriel's desire to create a photographic sleeve with the qualities of a painting. Storm Thorgerson came up with the design after having a dream in which Peter Gabriel's face was melting, as if he were in a horror film; Thorgerson liked the idea of mixing real life with dream life on the sleeve of Gabriel's next album. 'OK, I can handle that – I can handle being in your dreams so long as you are not in mine,' Gabriel told him. The melting effect was achieved by rubbing a freshly taken Polaroid photograph as the chemicals developed, manipulating their positions before the chemicals cooled and settled, to create the eerie, misshapen image. It gives the feel of there being something underneath the surface, which is apt for an album that is Gabriel's strongest, the one where his musical and thematic ideas meshed perfectly, with substance and surface combining to make an intrinsic whole.

RECORD LABELS

Charisma, Mercury

RELEASE DATES

May 1980

SONGWRITERS

Peter Gabriel

peter
gabriel

AC/DC
BACK IN BLACK (1980)

Shortly before recording sessions for the **Back In Black** album began in April 1980, AC/DC's original lead singer, Bon Scott, died of alcohol poisoning. As such, when the album was ready for release, the band decided to put it in an all-black sleeve. Lead guitarist Angus Young has confirmed that it was 'as a sign of mourning' for Scott, while the album's title reflects this and the fact that AC/DC considered splitting up before recording the album with new lead singer Brian Johnson. As a simple statement it works much in the same way as the Beatles' **The Beatles** (see page 63), and has become one of the most recognizable album sleeves of all time, thanks in no small part to its crunching title track. Of course, its image also lives on in the public's mind due to its influence on the sleeve for **Smell The Glove** by spoof heavy-metal band Spinal Tap. Forced to censor their original artwork, which featured a nude lady on all fours with a dog collar around her neck being forced to smell a glove, Spinal Tap's lead guitarist Nigel Tufnel describes the revised plain black artwork as, 'How much more black could this be? And the answer is none. None more black.'

RECORD LABELS

Atlantic

RELEASE DATES

July 1980

SONGWRITERS

Brian Johnson, Angus Young, Malcolm Young

AC/DC

BACK IN BLACK

CAPTAIN BEEFHEART & THE MAGIC BAND

DOC AT THE RADAR STATION (1980)

Since blowing everyone's minds with **Trout Mask Replica** (1969), still one of the most misunderstood – and hardest to understand – albums of all time, all eyes were on John Peel favourite Captain Beefheart (real name Don Van Vliet) to further push the sonic boundaries. The mid-70s saw him hit a relatively infertile period, however, but by the early 80s he seemed to have regained his creativity. Some might suggest this was because **Doc At The Radar Station** relied in places on reworked versions of some material from the **Trout Mask...** sessions, but it is also fair to note that Vliet had started painting more around this time (and would later retire from music entirely to concentrate more on his painting), and had probably been enthused by this new creative outlet. Like **Trout Mask...**, the **Doc...** sleeve plays with masking and identity, with one darkly shadowed character almost unidentifiable on the right-hand-side. It was a precursor of the style of abstract art that Vliet would develop later on and, as the penultimate Captain Beefheart album, perhaps helped Vliet to realize where his creative future lay.

RECORD LABELS

Virgin

RELEASE DATES

August 1980

SONGWRITERS

Don Van Vliet

DOC AT THE RADAR STATION CAPTAIN BEEFHEART AND THE MAGIC BAND
VAN VLIET © 77

DEAD KENNEDYS

FRESH FRUIT FOR ROTTING VEGETABLES (1980)

Dead Kennedys were one of the earliest, and still most revered, American hardcore punk bands, sounding something like a rockabilly Ramones (only a hundred times faster) and helping to point the way towards psychobilly. The sleeve for **Fresh Fruit For Rotting Vegetables** mirrors the 'raze and burn' sound of the music, using a striking, politically charged image, which no doubt later inspired Rage Against The Machine's debut album sleeve (see page 305). Introducing their similarly unapologetic left-wing politics, the Kennedys chose a black-and-white photo of police cars set on fire during the White Night Riots of 21 May 1979. The riots began as a peaceful demonstration from San Francisco's gay community, as they protested against the lenient sentence given to San Francisco City Advisor Dan White, who had murdered Mayor George Moscone and his homosexual Supervisor Harvey Milk. In fact, the rear sleeve – featuring a doctored image of Sounds Of Sunshine, an old lounge act – caused more controversy. The Kennedys pasted their logos over the band's equipment, but vocalist Warner Wilder threatened to sue. The band proposed a compromise by removing the Sounds Of Sunshine band members' heads in the image, but eventually had to settle for a photo of four old ladies sitting in a front room.

RECORD LABELS

Cherry Red, Faulty-IRS

RELEASE DATES

September 1980 (US: November 1980)

SONGWRITERS

Principal songwriter: Jello Biafra

Secondary songwriters: 6025 (real name Carlos Cardona), East Bay Ray, John Greenway, Doc Pomus/Mort Shuman

PRINCE

DIRTY MIND (1980)

Up until **Dirty Mind**, Prince had been hailed as the new Stevie Wonder – an R&B singer who played all his own instruments. With his third album, however, he did a complete about-face, stepping out in a look that announced, 'The sex-obsessed pixie has landed'. It shocked those who expected more R&B ballads and pop funk aimed at the dancefloor. According to Prince's then-guitarist Dez Dickerson, up to this point Prince had worn gold lamé leggings onstage, with no underwear underneath. When his manager baulked, telling Prince that he had to wear underwear, he took it quite literally and stepped out in a flasher's mac with only women's bikini bottoms and stockings on underneath. 'I'm going to portray pure sex' was Prince's explanation; it certainly wasn't the boy-next-door look. Even the upturned bed behind him, springs exposed, spelled an almost cell-like danger. Prince took a more stripped-down, new-wave direction for this album, and writ it large on the sleeve, all the way down to the 2-Tone 'Rude Boy' badge. Only the bravest fancy-dress partygoers would consider this as an option.

RECORD LABELS

Warner Bros.

RELEASE DATES

October 1980

SONGWRITERS

Principal songwriter: Prince

Secondary songwriters: Matt Fink, Morris Day

TALKING HEADS

REMAIN IN LIGHT (1980)

In keeping with the long line of Talking Heads album covers that were designed to distort the band's real image, the **Remain In Light** sleeve is actually more revealing than any of their previous album sleeves – if you know what the signs mean. Budapest-born art designer Tibor Kalman designed the rear sleeve, which depicted a fleet of fighter planes seemingly shot with an infrared camera, which was a reference to bassist Tina Weymouth being the daughter of a United States Air Force General. It's not much, but it is a start. The front sleeve, also designed by Kalman, goes back to the usual Heads image-distorting theme. A Beatles' **Let It Be**-style design, with each band member's face electronically altered, points the way towards the more electro-styled music that the group had worked up for the album. It also reflects the somewhat distancing nature of David Byrne's lyrics, which revel in disorientation, irony and distrust. Interestingly, the use of upside-down 'A's in the band's name was an influence on the Nine Inch Nails' 'NIN' logo, which reverses the second 'N'.

RECORD LABELS

Sire

RELEASE DATES

October 1980

SONGWRITERS

David Byrne, Brian Eno, Chris Frantz, Jerry Harrison, Tina Weymouth

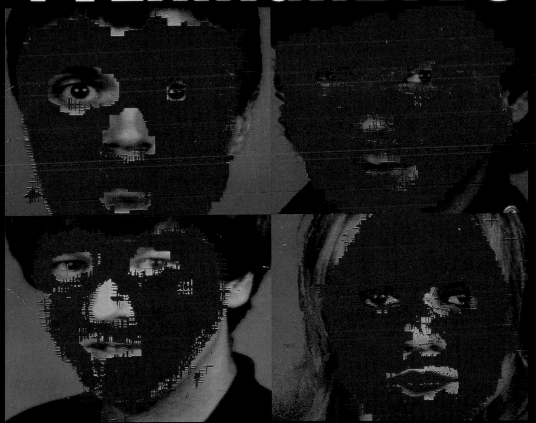

ADAM & THE ANTS

KINGS OF THE WILD FRONTIER (1980)

In the 80s a musician's image was more defining than anything else – often even more so than the music. You will remember Flock Of Seagulls for their haircuts, for instance, but can you remember any of their songs? Some artists managed to mix image and music perfectly, and Adam Ant continues to live on as one of the most successful to make the transition from new wave to new romantic, keeping both his look and sound fresh. For many thirtysomethings in the 00s, **Kings Of The Wild Frontier** will be the first album they ever bought, its blurred shot of Adam Ant on the road to becoming the highwayman of 'Stand And Deliver' promising raw rock'n'roll energy, while the bright colours announced Adam Ant as a bang-up-to-date pop star for the 80s teen to idolize. A flavour of Ant's electro-pop music came through with the colours, while the look – a mix of flamboyant historical clothing, merging Victorian-styled fashions and glam-rock make-up – defined British pop bands throughout the early 80s. With his major label debut, Adam Ant made sure he was there first.

RECORD LABELS

CBS, Columbia

RELEASE DATES

November 1980 (US: February 1981)

SONGWRITERS

Adam Ant, Marco Pirroni

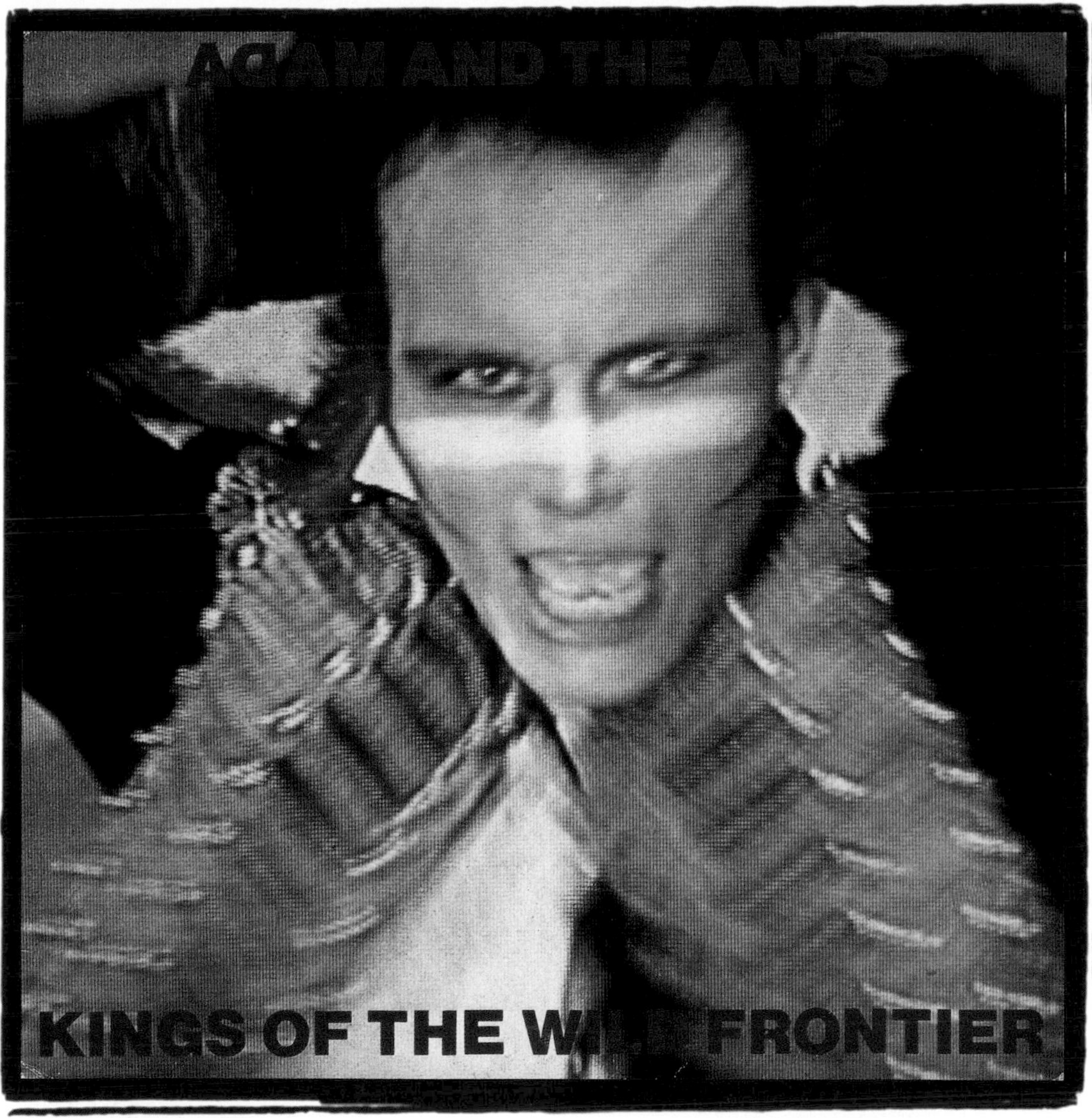
AND THE
KINGS OF THE
FRONTIER

X

LOS ANGELES (1980)

Though they formed in 1977, X didn't release an album until 1980's Ray Manzerek-produced **Los Angeles**. The record proved so influential upon Los Angeles musicians of the day (the group didn't restrict themselves to punk, but also dealt in rockabilly, ballads and a country influence) that X were later awarded an Official Certificate of Recognition from the City of Los Angeles, acknowledging their contributions to the local music scene and culture. In keeping with the stark, black-and-white photographic imagery of other US hardcore bands at the time (see Dead Kennedys' **Fresh Fruit For Rotting Vegetables**, page 219), X's **Los Angeles** sleeve is as iconic and striking as it is borderline offensive. A simple, burning wooden 'X' is an obvious ground zero-style manifesto from the position of Los Angeles punk, but also mimics the burning cross symbolism often associated with the Ku Klux Klan. There is no suggestion whatsoever that X had anything to do with KKK sentiments (in fact, very much the opposite), but as far as arresting images go, the burning cross is one of the more inflammatory (no pun intended). Considering the fact that Los Angeles would be set ablaze by race riots just 12 years later, it is also remarkably prescient.

RECORD LABELS

Slash

RELEASE DATES

1980

SONGWRITERS

Principal songwriters: John Doe/Exene

Secondary songwriters: John Densmore, Robbie Krieger, Ray Manzerek, Jim Morrison

LOS ANGELES

ELVIS COSTELLO & THE ATTRACTIONS

TRUST (1981)

Would you buy a used car from this man? For three albums after 1978's **This Year's Model**, Elvis Costello hadn't appeared clearly on an album sleeve until **Trust**. When he did, though, it wasn't quite with the same youthful verve of his initial new-wave look. Here Costello wears a tinted take on his NHS specs, looking over the top, eyebrows raised in a somewhat cynical, quizzing manner. Do you trust him? Or can he trust you? With hindsight it is easy to see where such cynicism comes from. There were tensions within the Attractions, and Costello's marriage to Mary Burgoyne was beginning to show cracks. Not only that, Costello, a keen chronicler of English life, was less than pleased with the election of the Conservative government, leading to a general sense of disillusionment. Looking for a safe haven, it is ironic that Costello's fanbase should, at the same time, be forced to ask whether they can trust him. If his previous album **Get Happy!!** can be seen as the party, being a soul music *tour de force*, then **Trust** is the comedown, the eclectic mish-mash covering jazz, rockabilly and pop. So which way would he turn next, and could you trust that it wouldn't lead you down a blind alley?

RECORD LABELS

F-Beat, Columbia

RELEASE DATES

January 1981

SONGWRITERS

Elvis Costello

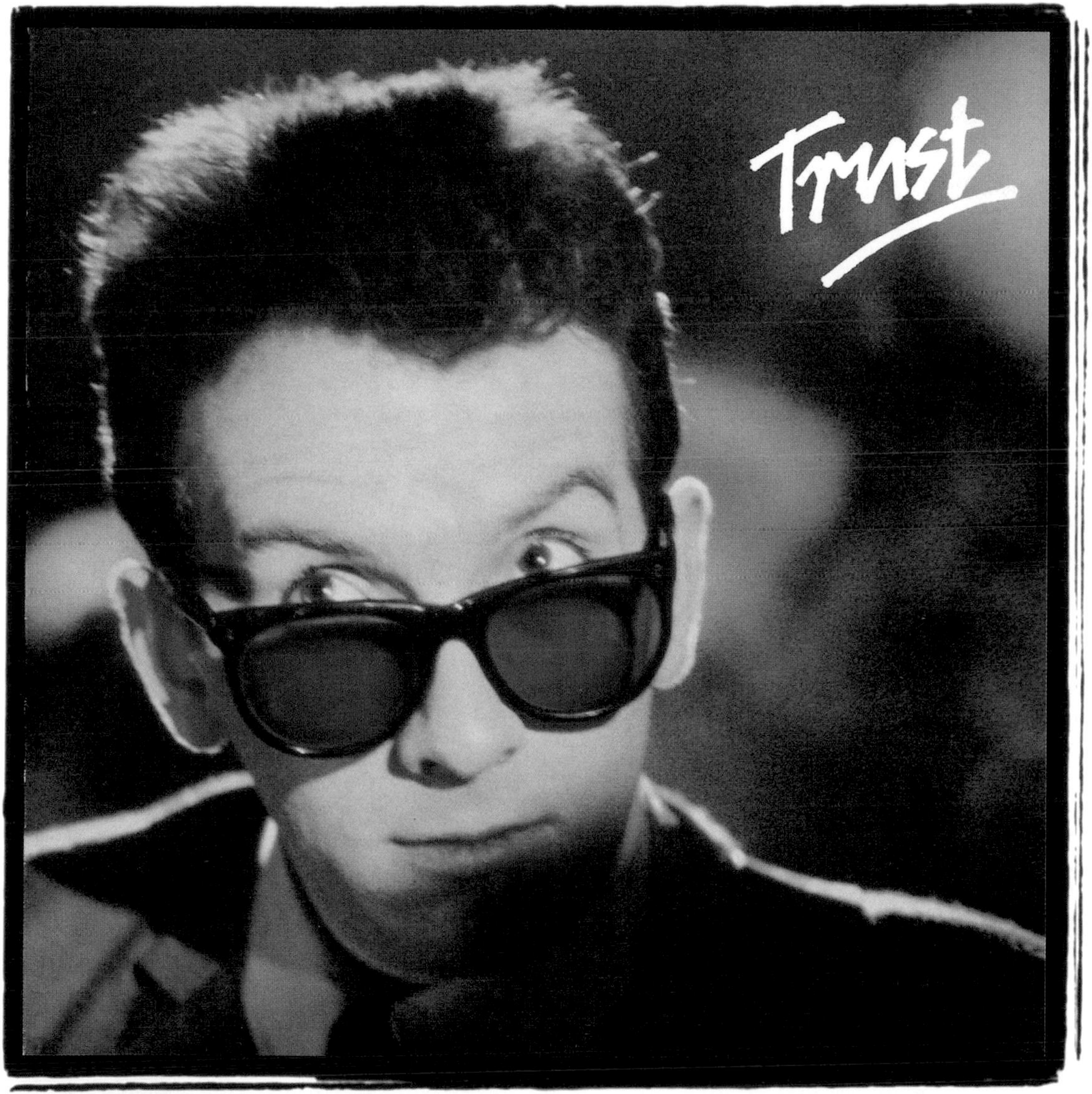
Trust

DEBBIE HARRY

KOO KOO (1981)

Blondie had long mixed disco, new wave and power pop together with hits such as 'Rapture'. **Koo Koo** took this forward even further, merging rock and funk and somewhat paving the way for producer Nile Rodgers' work on David Bowie's **Let's Dance** in 1983. For the most part, however, H.R. Giger's sleeve attracted most of the attention, and saw Debbie Harry move away from her established blonde-bombshell image in Blondie, even though she was still in the band when she released this album. Giger had already designed sleeves for Emerson, Lake & Palmer (see **Brain Salad Surgery**, page 137) and by 1981 was famed for his designs on the *Alien* film. For the **Koo Koo** sleeve he said, 'Since I had just had an acupuncture treatment … the idea of the four needles came to me, in which I saw symbols of the four elements, to be combined with [Harry's] face.' The shocking image inspired a promotional campaign which would have seen large-scale posters of the sleeve put up in London Underground stations, though it was ultimately decided that it was too disturbing to be plastered across public transport routes.

RECORD LABELS

Chrysalis

RELEASE DATES

August 1981

SONGWRITERS

Bernard Edwards, Debbie Harry, Nile Rodgers, Chris Stein

BOW WOW WOW

SEE JUNGLE! SEE JUNGLE! GO JOIN YOUR GANG, YEAH. CITY ALL OVER! GO APE CRAZY! (1981)

With Bow Wow Wow Malcolm McLaren proved that he still knew how to court controversy, after turning the world upside down with the Sex Pistols. Whereas Johnny Rotten was his Pistols pin-up, this time around he made the focus 15-year-old Annabella Lwin, and convinced Adam Ant's then-backing band to leave Ant and join Bow Wow Wow.

See Jungle!... is the kind of sleeve you just would not be allowed to get away with these days, and it caused no shortage of outrage in 1981. The sleeve was based on Édouard Manet's painting *Le Déjeuner Sur L'Herbe (The Luncheon On The Grass)*, which itself caused great offence in the 1860s, as it featured a naked woman sitting and picnicking with a group of men. By remaking the sleeve with Annabella taking the woman's part, Malcolm ensured the album would get noticed. Annabella's body was actually facing away from the camera in the sleeve, but it still attracted accusations of child pornography, while in America a different photo had to be used, which featured Lwin covered in a see-through dress. Lwin's mother tried to have the photo stopped altogether, but McLaren eventually won out.

RECORD LABELS

RCA

RELEASE DATES

October 1981

SONGWRITERS

Matthew Ashman, Dave Barbarossa, Leigh Gorman

THE POLICE

GHOST IN THE MACHINE (1981)

Though breaking big on the crest of the new-wave wave, the Police were really just riding a wagon while they turned out more sophisticated stadium pop rock that no true new-wave act would ever come up with. It is no surprise, then, that the sleeve for their fourth album is an equal mix of new-wave imagery and more commercial concerns. Though it is as starkly designed as Joy Division's **Unknown Pleasures** (see page 197), compare the image with that of Joy Division's pulsar radio wave and it is nowhere near as oblique. LED displays were becoming big business in the early 80s, most notably on digital clocks, and the Police's use of such an instantly recognizable image for **Ghost In The Machine** is a far cry from radio waves, and much more commercially savvy. Everyone would recognize the bright red lines against a black background, and anyone with just a passing interest in the band would realize that the three figures represented the band (Andy Summers, left; Sting, centre; Stewart Copeland, right).

RECORD LABELS

A&M

RELEASE DATES

October 1981

SONGWRITERS

Principal songwriter: Sting

Secondary songwriters: Stewart Copeland, Andy Summers

THE **POLICE**

GHOST IN THE MACHINE

IRON MAIDEN

THE NUMBER OF THE BEAST (1982)

Iron Maiden's third album became the first heavy-metal LP to reach No. 1 in the UK charts, and it is not going too far to suggest that by this point their image was a large part of their popularity. Since the beginning artist Derek Riggs had furnished Maiden's artwork with incredibly detailed paintings of a character known as Eddie. The Eddie image, based on an old Riggs painting of a fictional punk, lives with Maiden to this day, though his appearance on **The Number Of The Beast** is notable because it is the first time Eddie appears out of a normal 'street' context. The previous two sleeves saw Eddie generally looking like a delinquent on the street, while later ones would see him in space, in a mental institution, taking on super powers and forming part of the Egyptian Pyramids. Here he is about to crush the Devil on an artwork originally intended to grace the picture sleeve of Maiden's 'Purgatory' single. When the band saw it, they thought it was too good for a single release and decided to use it for **...Beast**, asking Riggs to design something simpler for 'Purgatory'.

RECORD LABELS

EMI, Harvest

RELEASE DATES

March 1982

SONGWRITERS

Clive Burr, Paul Di'Anno, Steve Harris, Adrian Smith

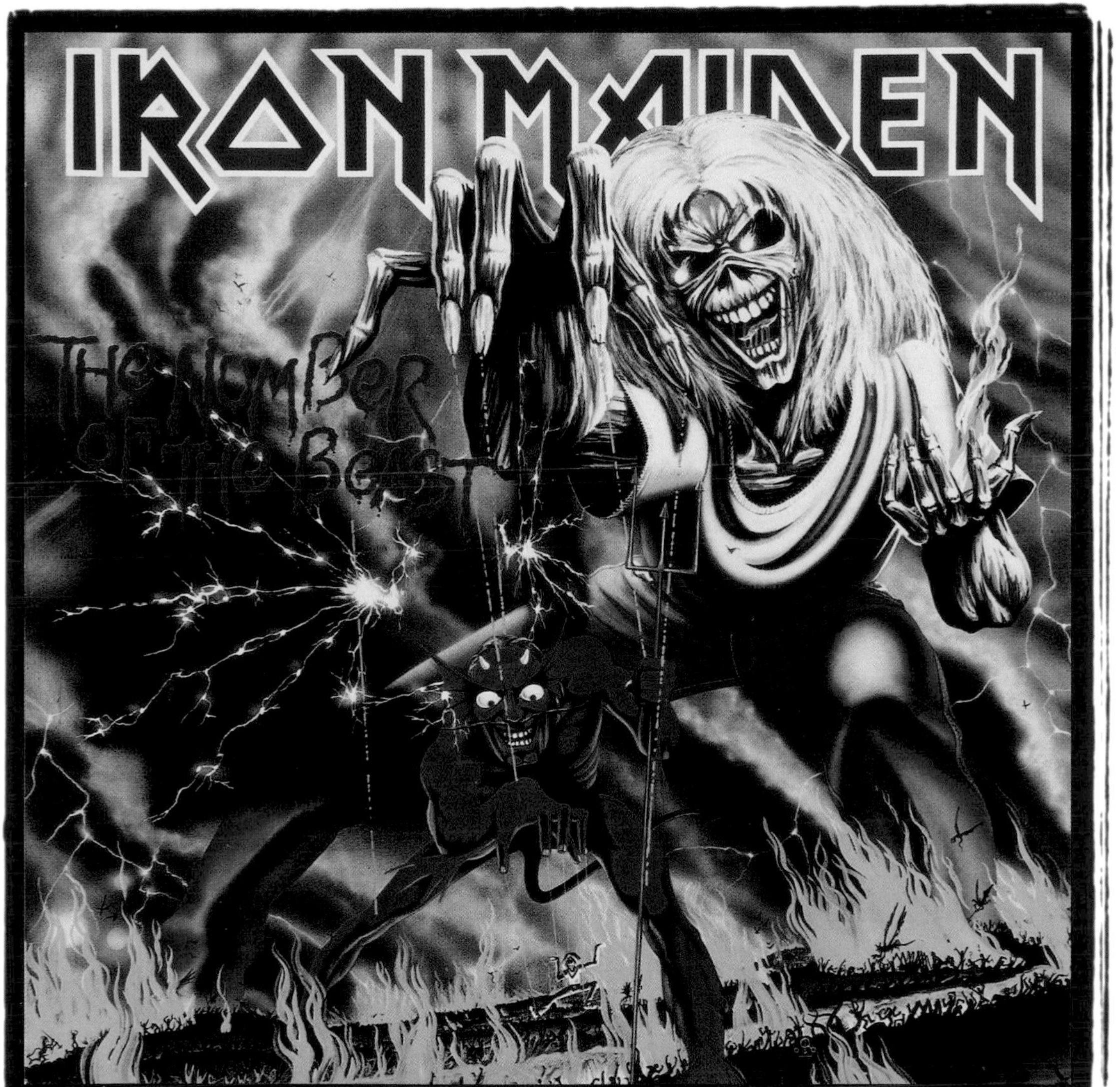
IRON MAIDEN
The Number of the Beast

TALKING HEADS

THE NAME OF THIS BAND IS TALKING HEADS (1982)

Talking Heads' first live album chronicled the growth of the band over their first four studio albums, both physically and artistically. A double album, it essentially charts the group's greatest hits from their debut **Talking Heads: 77** to 1980's **Remain In Light** (see page 222). The album's artwork featured David Hockney-style joiner photographs that first capture the original four-piece playing live in a front room, continuing across the back sleeve and the two inner picture sleeves, presenting the growing band, until finishing with the 10-piece group that toured in 1981. With individual artist shots interspersed throughout, it is almost a scrapbook of Talking Heads throughout their most influential phase. For a band that so often eradicated their actual likenesses on album sleeves, it is fitting that they reveal themselves in all their sprawling glory on a live recording. In fact, it is the only Heads artwork that fully reveals the group. Compare this with the sleeve for 1984's live album **Stop Making Sense**, and it seems that Talking Heads circa 1982 were almost proud of their achievements, while **Stop Making Sense** very much adheres to the later stereotype of 'big-suited' David Byrne, without even showing his face.

RECORD LABELS

Sire

RELEASE DATES

April 1982

SONGWRITERS

Hugo Ball, David Byrne, Brian Eno, Chris Frantz, Jerry Harrison, Tina Weymouth, Wayne Zieve

THE NAME OF THIS BAND IS TALKING HEADS

1982 SIRE RECORDS. DISTRIBUTED BY WEA RECORDS LTD ® A WARNER COMMUNICATIONS COMPANY. COUNTRY OF MANUFACTURE OF RECORD AS STATED ON RECORD LABEL. PRINTED AND MADE IN ENGLAND.

I ZIMBRA   DRUGS
HOUSES IN MOTION   LIFE DURING WARTIME

LIVE 2 LP SET (1980-1981)

TAKE ME TO THE RIVER
THE GREAT CURVE   CROSSEYED AND PAINLESS

DURAN DURAN

Rio (1982)

Duran Duran were the epitome of the new romantic bands that relied as heavily on their image as they did on their music. The video for 'Rio' alone is as much an expensive holiday advert as it is a pop video, and it seemed that, in Duran Duran's world, everything had to be perfect on the surface before it could be perfect anywhere else. For **Rio**'s cover, the decadent, jet-set lifestyle that they symbolized is mirrored by an artwork painted by Patrick Nagel, an artist noted for his designs of women in an Art Deco style. His approach was to work from a photograph and remove any parts of an image that he felt were superfluous, lending his art a two-dimensional immediacy that recalls Japanese woodblock prints and some of Andy Warhol's screen prints. Oddly, the sleeve used for **Rio**'s original release is of a slightly more sinister-looking portrait than another artwork Nagel submitted for the band's consideration. The second image, a sexier, more suggestive shot of a woman on a bed, was used for the 2001 CD reissue of **Rio**.

RECORD LABELS

EMI, Capitol

RELEASE DATES

May 1982 (US: January 1983)

SONGWRITERS

Simon Le Bon, Nick Rhodes, Andy Taylor, John Taylor, Roger Taylor

DEPECHE MODE

A BROKEN FRAME (1982)

Everything about **A Broken Frame** suggests hard graft from the band that would become one of the world's most noted electro pioneers. Their debut, **Speak And Spell** (1981), may have been more of a straight disco-pop album, but it gained critical ground that gave the band a firm foundation. So, when chief songwriter Vince Clarke left Depeche Mode after its release, he also left them with a hole that desperately needed filling. Martin Gore took over the duties full-time for **A Broken Frame**, and though he would continue to be chief songwriter for a while, it is odd that many of the sonic experiments on **A Broken Frame** were not taken up by the band on later albums; the record seems to have fallen through the cracks in Depeche Mode's history. As the sleeve suggests, it was difficult trying to stay standing after Clarke's departure, and though some of Gore's ideas weren't developed, **A Broken Frame** is the sound of a group working hard to find out what they could and could not do without their main songwriter. What the sleeve doesn't show is that they reaped the rewards of experimentation; sometimes it is just as handy knowing what you don't want to do as knowing what you do.

RECORD LABELS

Mute, Sire

RELEASE DATES

September 1982

SONGWRITERS

Martin Gore

U2

WAR (1983)

War saw U2 fully step into the protest-rock shoes that they, and most notably Bono, have now long been known for filling. Opening with 'Sunday Bloody Sunday', the album was immediately pegged as the group's most overtly political record to date. Bono has said that, as war raged in the Falklands and the Middle East, 'By calling the album **War** we're giving people a slap in the face and at the same time getting away from the cosy image a lot of people have of U2.' Indeed, the only constant in their image here was using Peter Rowan on the sleeve again, the brother of Bono's friend Guggi Rowan. Peter had appeared on the sleeve of U2's debut **Boy** and would later appear on the sleeves for a greatest hits, their **Three** EP and a collection of demo recordings; having such a young face stare out so accusingly, with the band's name and title emblazoned in bright red down the side of the sleeve, forced the buyer to actually stop and think when buying this pop record. U2 would continue to mine this 'stop and think' seam throughout their career, but **War** was the album that set their political-rock stall out.

RECORD LABELS

Island

RELEASE DATES

February 1983

SONGWRITERS

Bono, Adam Clayton, Edge, Larry Mullen Jr.

ZZ TOP

ELIMINATOR (1983)

Eight albums in, **Eliminator** saw ZZ Top go stratospheric, selling enough copies to be certified Platinum ten times over and reaching Diamond sales status. ZZ Top frontman Billy Gibbons was a hot rod obsessive, and his 'Eliminator' hot rod (as it is now known) is probably one of the most famous cars of its type in the world. The 1933 Ford Coupe was custom-built for Gibbons in the early 80s, and featured the ZZ Top logo (on the top of the album sleeve) painted down the side of the fire engine-red car. That a painting of the Ford Coupe appeared on the sleeve of **Eliminator** probably did more for the original cars (previously immortalized by Brian Wilson as a drag racer's choice of vehicle in the Beach Boys' song 'Little Deuce Coupe') than a lifetime's worth of advertising could. The car even featured in a number of ZZ Top promotional videos, which introduced a new generation of youths to a vehicle that had already played a major part in the history of America's automobiles. In these, the Eliminator was usually a symbol of fantasy escapism that came along to rescue luckless teenage boys, as it arrived full of sexy ladies to turn their lives around (as in the video for 'Legs').

RECORD LABELS

Warner Bros.

RELEASE DATES

June 1983 (US: April 1983)

SONGWRITERS

Frank Beard, Billy Gibbons, Dusty Hill

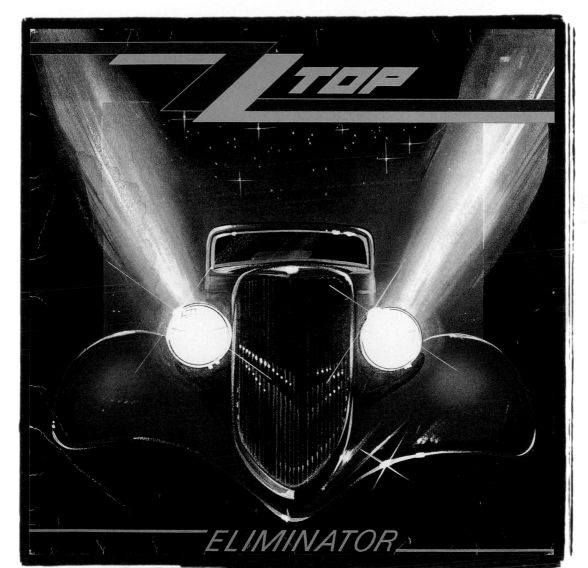

NEW ORDER

POWER, CORRUPTION AND LIES (1983)

Though actually New Order's second album, **Power, Corruption And Lies** saw the band really shake loose the Joy Division shackles (they were the surviving members, plus Gillian Gilbert, having opted to stay together after Joy Division's lead singer Ian Curtis committed suicide). Like AC/DC's **Back In Black** (see page 215), the album can be seen as a commemorative sleeve in honour of the band's departed lead singer, while Joy Division/New Order artist Peter Saville wanted to make something that directly juxtaposed romanticism with more processed, modern imagery. Hence the band's name and album's title were represented by a colour code in the top right-hand corner of the sleeve, with a decoder on the back (a similar code featured on New Order's 'Blue Monday' 12-inch, which is still the best-selling 12-inch of all time and, so the legend goes, was so expensive to make that the band lost money every time a copy was sold). The National Heritage Trust allegedly refused Factory the right to use the painting, which was by Henri Fantin-Latour. Factory boss Tony Wilson is said to have asked the Trust who actually owned it. When they replied that it was, technically, the British public, Wilson allegedly told them that 'the British people now want it', and went ahead.

RECORD LABELS

Factory, Streetwise

RELEASE DATES

May 1983

SONGWRITERS

Gillian Gilbert, Peter Hook, Stephen Morris, Bernard Sumner

CYNDI LAUPER

SHE'S SO UNUSUAL (1983)

What Lauper lacked in true musical edge, she certainly possessed in visual impact. In the era when MTV was young, but fast becoming a dominant force (one video on heavy rotation could make you a star; look what it did for Run-DMC or Prince, whose 'When You Were Mine' Lauper covers here), the image was almost more important than the music to some artists, whose sole goal was to make it big. What we got with **She's So Unusual** is essentially the 80s' overriding fashion for girls for the rest of the decade – even more ahead of the curve than Madonna's look in *Desperately Seeking Susan*, which wasn't released until the following year. Think of any 80s teen flick and you've got the look: second-hand dresses and baggy tops thrown together with bangles and fishnet stockings, all in gaudy Day-Glo colours. Lauper was so era-defining that her music was prominently featured in *The Goonies*, released the following year, and is as much a landmark for children of the 80s as the Lauper look or *Ghostbusters* (1984).

RECORD LABELS

Portrait

RELEASE DATES

December 1983

SONGWRITERS

Principal songwriters: Rick Chertoff, Gary Corbett, Rob Hyman, Cyndi Lauper, Stephen Broughton Lunt, Jules Shear, John Turi

Secondary songwriters: Tom Gray, Prince, H. Huss/Mikael Rickfors, Al Lewis/Al Sherman/Abner Silber

VAN HALEN

1984/MCMLXXXIV (1984)

It was all too easy for heavy-metal groups to not quite see the painful irony of things – mostly their image and their artworks. In 1984, things got a little George Orwell-*1984*, and in the hands of other bands, god only knows what horrific Big Brother-style sleeves would have been designed for an album of the same name in the interests of really hammering a point home. The sleeve for **1984** (annoyingly written in Roman numerals for some smart-assed reason) is pretty much a masterpiece, however. The baby angel with a pack of cigarettes gives the record a real edge, despite the fact that it is such a softly focused painting. Sadly, however, the great design didn't stretch to the rest of the album; maybe they had blown their budget. On the inner and rear sleeves things do succumb to the *1984* cliché: a terrible futuristic band image with the early computer-style '1984' font, and a black-and-white photo of what looks like an aerial shot of a bed of CCTV cameras, presumably intended to give a 'You are being watched' feel. A great way to ruin a great look.

RECORD LABELS

Warner Bros.

RELEASE DATES

January 1984

SONGWRITERS

Principal songwriters: Michael Anthony, David Lee Roth, Alex Van Halen, Eddie Van Halen

Secondary songwriter: Michael McDonald

THE SMITHS

THE SMITHS (1984)

As time wore on the Smiths would become as known for their album and single picture sleeves (they refused to appear on any of their own sleeves in the UK) as they would for their life-changing music and being the world's biggest indie band. Their sleeves were always iconic images, usually of lesser-known movie and pop stars as picked by Morrissey, and then cut and cropped into the ideal image by Jo Slee, the Rough Trade art director. For their first album, however, the band were still an unknown quantity, and back when **The Smiths** was released, not putting yourself on the sleeve when you were a new band looking to make it big was something of a left-of-centre move. Here, Morrissey instead chose a still from Andy Warhol's film *Flesh*, made in 1968. The actor in the photo was actually Joe Dallesandro, who is alleged to be the model on the Rolling Stones' **Sticky Fingers** sleeve (see page 99), and avid music fan Morrissey is sure to have known this. What it presented to the world, though, was a subtle mix of masculinity and delicacy, a path the Smiths would walk along time and time again.

RECORD LABELS

Rough Trade, Sire

RELEASE DATES

February 1984

SONGWRITERS

Johnny Marr, Morrissey

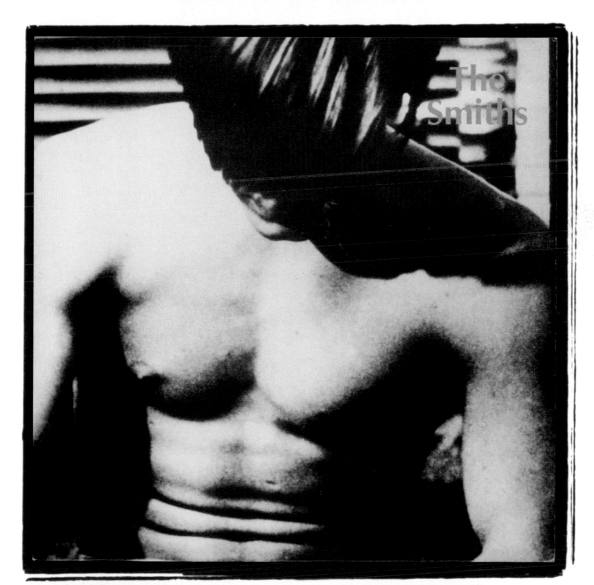

The
Smiths

BRUCE SPRINGSTEEN

BORN IN THE USA (1984)

If anyone still questioned Springsteen's ability to speak to everyone (even the British, so far removed from the Boss's blue-collar America), **Born In The USA** came along to silence them. In 1984 Springsteen went from having a strong fanbase to being a global icon. From the hugely radio-friendly sound of the music down to the image on the front cover, **Born In The USA** was an exercise in making the medicine go down sweetly. The album's singles were Springsteen's most bombastic recordings yet, while the title track's anthemic chorus led many to miss the song's deeper meaning. At a glance, however, it is easy to see **Born In The USA** as a generally pro-American album. The slack denim jeans and cap lazily thrown in the back pocket still proved Springsteen to be the voice of the common man, but the US flag in the background was all too quickly picked up as a patriotic gesture, as opposed to a despairing look from the disillusioned millions. Still, this was the year when Ronald Regan won an almost landslide presidential election, so it is fair to assume that right-wing patriotism was at an all-time high. That anyone's conscience would dare say a thing against the motherland was surely unthinkable.…

RECORD LABELS

CBS, Columbia

RELEASE DATES

June 1984

SONGWRITERS

Bruce Springsteen

BORN IN THE U.S.A./BRUCE SPRINGSTEEN

THE SMITHS

MEAT IS MURDER (1985)

For their second album proper the Smiths proved that a strong image was their stock-in-trade, and if anyone misunderstood the title track itself, here it is in all its black-white-and-green glory on the sleeve. Morrissey can be an unsubtle man when he wants to, and here the point is as on-the-nose as you like: whether it be animal or human, a life is still a life and eating meat is comparable to the murder of innocent youths. It didn't do the band any favours, as some critics had already blamed Morrissey for courting controversy with 'Suffer Little Children' from the Smiths' debut album and, as one reviewer said, the 'simulated bovine cries and buzz-saw guitars, takes vegetarianism to new heights of hysterical carniphobia'. Still, this is only one year after Benetton hired art director Oliviero Toscani to oversee its marketing campaigns. Toscani would come up with some of the most controversial advertising images in history, making **Meat Is Murder** rather tame in comparison. Morrissey was serious about the image, though, and rumour has it that he forbade the rest of the band to be photographed eating meat.

RECORD LABELS

Rough Trade, Sire

RELEASE DATES

February 1985

SONGWRITERS

Johnny Marr, Morrissey

GRACE JONES

SLAVE TO THE RHYTHM (1985)

After three years away from recording, Grace Jones knew how to show everyone she had returned. Where most of her past album sleeves were rather po-faced, **Slave To The Rhythm** picks up the Art-Deco influence and playful approach to her angular features that came in with its predecessor, **Living My Life** (1982). The image is as in-your-face as the album, which is built heavily around remixes of the title track and a collection of spoken-word pieces about Grace Jones's life. The clash of sounds, including heavy beats courtesy of Trevor Horn and an almost eardrum-shattering vocal from Jones, resulted in Jones's most commercially successful LP in the UK, one that is well illustrated in the glass-shattering cover image. If this is the album equivalent of an audio biography, then, perhaps the sleeve is setting the record straight, too. It pretty much reconfigures Jones's image as someone much more approachable and full of humour than previously thought. The 'Kodak safe' tag placed in her hair like an afro comb raises questions, though – was that some sort of message to a photographer who crossed her path during her modelling career?

RECORD LABELS

ZTT-Island, Manhattan

RELEASE DATES

October 1985 (US: November 1985)

SONGWRITERS

Simon Darlow, Trevor Horn, Steve Lipson, Bruce Woolley

GRACE JONES SLAVE TO THE RHYTHM

MADONNA

TRUE BLUE (1986)

Across her previous two albums, Madonna's image had become increasingly sensual, and the photo on **True Blue**'s sleeve instantly became one of her most iconic poses as it shot around the world, reaching No. 1 in the album charts in 28 countries. This shot also shook off the slightly adolescent image Madonna had nurtured throughout 1984's **Like A Virgin** and her film role in 1985's *Desperately Seeking Susan*. American fashion photographer Herb Ritts took the shot, and the fact that he is best known for photographing portraits in a classical Greek style is not lost here: Madonna's name and pose all suggest a link to classical civilization. The colour shades work perfectly in matching the image to the title, and the whole is really an exercise in branding, to the point where each one of Madonna's album sleeves, up to and including 2000's **Music**, would rely heavily on blue costumes and props. It wasn't her most sensual artwork, of course. In 1992 the 'Erotica' picture disc, featuring Madonna sucking on the big toe of an unseen recipient, was withdrawn.

RECORD LABELS

Sire

RELEASE DATES

July 1986

SONGWRITERS

Stephen Bray, Gardner Cole, Brian Elliot, Brice Gaitsch, Patrick Leonard, Madonna, Peter Rafelson

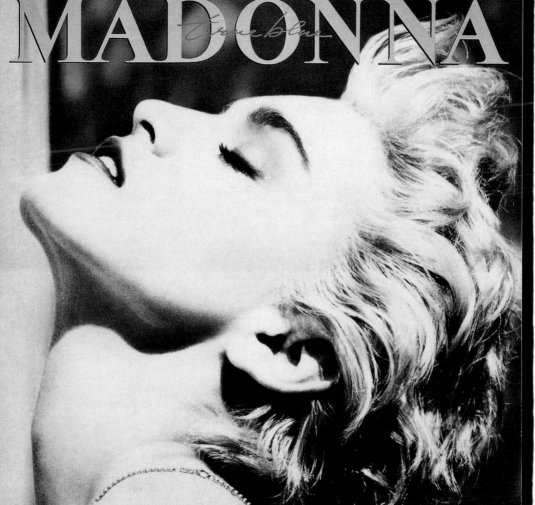

MILES DAVIS

TUTU (1986)

Oddly, for a man so obsessed with fashion and his appearance, Miles Davis saw his albums packaged in a string of gaudy sleeves in the 80s, with **You're Under Arrest** (1985) being possibly the most ridiculous. For his critical rebirth, and his first record for new label Warner Bros., Davis came out with something more fitting to his own great stature. Maybe the album's title dictated it: he had named it after Cape Town's first black Anglican Archbishop, and winner of the 1984 Nobel Peace Prize, Archbishop Desmond Tutu. The solemn, almost bust-like image of Miles' face, close-cropped and pressed right up to the camera, lends the album a certain weight. It is also another way of Miles, well known for his temper, getting in your face. Most of all, though, Miles probably recognized that **Tutu** was something of a milestone for him, and so wanted to give it a sleeve that wouldn't age in the same way as **You're Under Arrest** has. As such, it is presenting one of the most influential figures in jazz music, a man who had turned 60 earlier in the year, as a figure who still isn't going away, and who can continue to adapt and survive in jazz's shifting waters.

RECORD LABELS

Warner Bros.

RELEASE DATES

October 1986

SONGWRITERS

Principal songwriters: Miles Davis, Marcus Miller

Secondary songwriters: Davis Gamson/Green Gartside

BEASTIE BOYS

LICENSED TO ILL (1986)

Beastie Boys crash-landed quite literally on the sleeve of their debut, as the full gatefold sleeve was a painting of a Boeing 727 crashing into a mountain. But then on first listen, **Licensed To Ill** was the aural equivalent of being in an aeroplane crash. Always happy to needle people at the time, the '3MTA3' text on the side of the plane actually seems to spell out 'EATME' if you hold it up to the mirror (that's nothing – originally they wanted to call the album *Don't Be A Faggot*, but Def Jam refused to release it if they did). Run-DMC had pioneered the mix of rock and rap, but Beastie Boys did it with an understanding and humour not yet seen in hip hop. Like Run-DMC had made the transition to white audiences with their incorporation of rock music, Beasties were simply the first white hip-hop group that ever mattered to anyone. They might look back on some aspects of **License To Ill** with no small degree of awkwardness, but for everyone else this record was the start of a beautiful friendship.

RECORD LABELS

Def Jam

RELEASE DATES

November 1986

SONGWRITERS

Principal songwriters: Ad Rock, MCA, Mike D, Rick Rubin

Secondary songwriters: Darryl 'DMC' McDaniels, Joseph Simmons

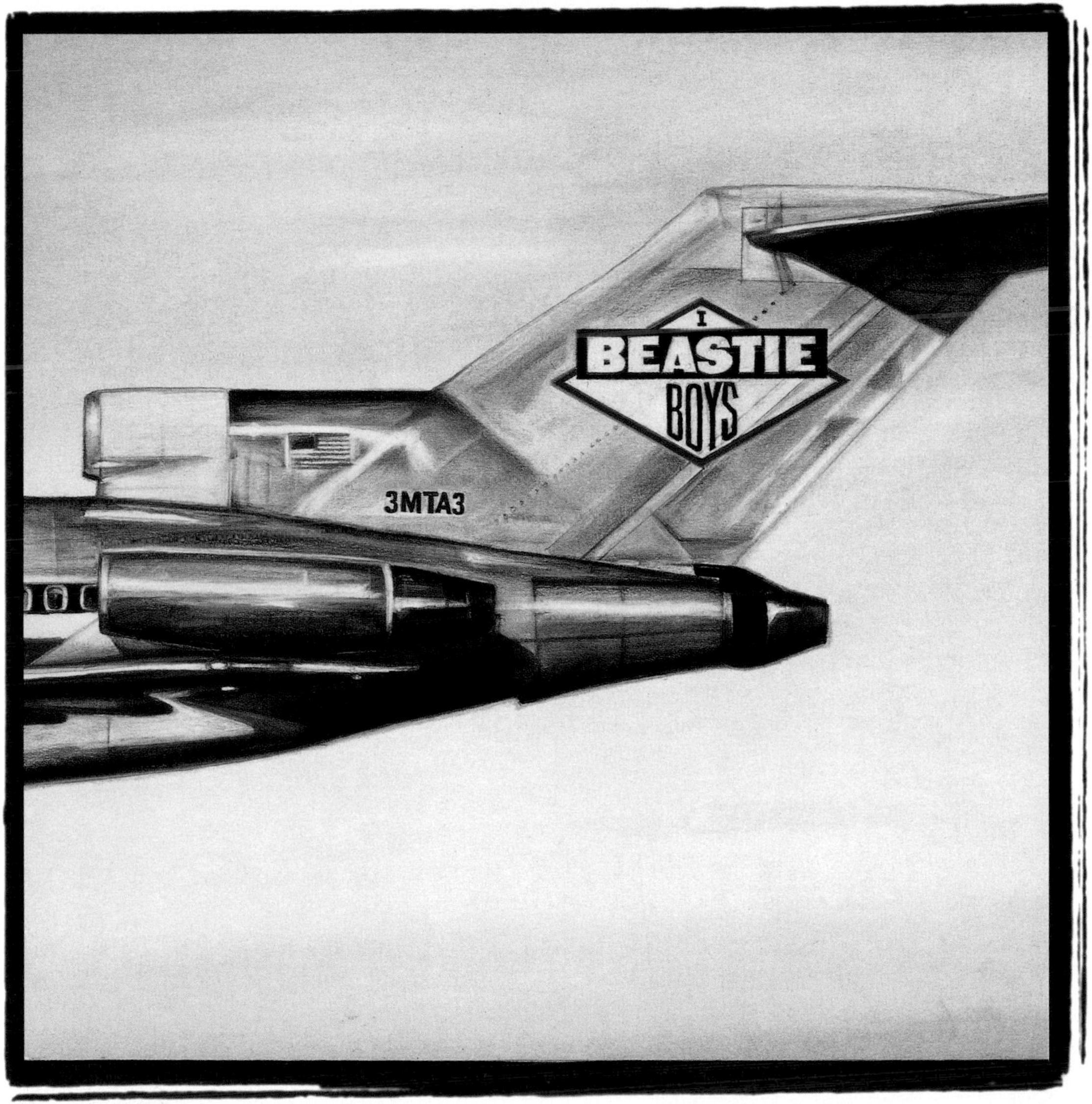
I
BEASTIE
BOYS
3MTA3

PRINCE

SIGN 'O' THE TIMES (1987)

Having sacked the Revolution, Prince set out to make his first album in four years that didn't credit anyone else. The sleeve sees a newly bespectacled Prince turning his back on the past, as he slips out of the bottom right-hand corner, almost undetected among the visual overload. The 'Girls Girls Girls' sign in the background suggests that he is walking away from his promiscuous lifestyle to settle down and tackle more serious issues. While this wasn't entirely true, many of the songs refer to monogamy for the first time on a Prince album, while the title track is concerned with a world in decline. The messy stage set-up suggests a new-age settlement and is almost a visual statement on the state of the world. Most telling, however, is Prince's first visual stab at his label Warner Bros. (during increasingly bitter arguments with them in the mid-90s, he would perform with 'Slave' written on his cheek). The crystal ball on the drum riser is a reference to the album Prince thought **Sign 'O' The Times** should have been: a triple-LP set entitled *Crystal Ball*. Warner Bros. refused to let Prince release such a lengthy product on commercial grounds, and thus the first link in a long chain of grievances was put in place.

RECORD LABELS

Paisley Park/Warner Bros.

RELEASE DATES

March 1987

SONGWRITERS

Prince

U2

THE JOSHUA TREE (1987)

Returning to **War**'s anthemic rock, **The Joshua Tree** wasn't as politically edged as that album, but instead saw U2 searching for solutions to their problems. The album's title was intended as a tribute to America, and the record was U2's most American-sounding to date, incorporating blues and country influences for the first time. When it comes to Americana, the Joshua tree is a weighty image, as country-rock hero Gram Parsons died aged just 26 at the Joshua Tree Inn. The trees themselves, however, are famous for their survival. 'It's supposed to be the oldest living organism in the desert', drummer Larry Mullen Jr. said about the band's choice to have their cover photo shot there. 'They can't put a time on it, because when you cut it, there's no rings to indicate how old it is. Maybe that's a good sign for the record.' Noted rock photographer Anton Corbijn took the shot, and in so doing created one of the most iconic album sleeves, not just for U2 but for the 80s. That their image sees them in an almost nonchalant pose, however, it further stresses U2's intentions for the album, as they essentially deconstructed America and put it back together in their image.

RECORD LABELS

Island

RELEASE DATES

March 1987

SONGWRITERS

Bono, Adam Clayton, Edge, Larry Mullen Jr.

TOM WAITS

FRANKS WILD YEARS (1987)

Franks Wild Years was adopted from the track of the same name on Waits' **Swordfishtrombones** album, and first saw light as a stage play co-written by Waits and his wife Kathleen Brennan. A romantic opera, it follows accordion player Frank's highs and lows as he chases fame, though the storyline is largely absent from the album. As a collection of songs, however, it is one of the high points in Waits' career. The glass is a bit heavy on the crushed dreams imagery, but the city skyline is a lovely detail placed across the accordion. At this time Waits was very taken by German music and an almost carnival approach to songwriting, and the **Franks Wild Years** sleeve is given a similarly European look. Of course, Waits had to present things slightly distorted from reality. Look closely and you will see that this is actually a photo of an accordion player with the head cut off, and a blurry Waits, strangely framed, stuck in its place. He would spend the entire 80s proving that he could make music from any found sounds and fuzzy instruments, and as his interest in the visual arts took off, Waits seemed to be proving he could do the same with images.

RECORD LABELS

Island

RELEASE DATES

August 1987

SONGWRITERS

Principal songwriter: Tom Waits

Secondary songwriters: Kathleen Brennan, Greg Cohen

TOM WAITS
FRANKS WILD YEARS

UN OPERACHI ROMANTICO IN TWO ACTS

MIDNIGHT OIL

DIESEL AND DUST (1987)

If it were not for **Diesel And Dust**'s heavy message being wrapped up in a more pop-music style, six albums in Australia's Midnight Oil may well have still been unheard of outside of their home country. As it was, **Diesel And Dust** made them global stars for a moment, and the group started to take on U2-like connotations of 'political alternative rock band' (also note the top and bottom border similarity between **Diesel And Dust** and U2's **The Joshua Tree**, see page 273, released the same year). They might have taken the weird eccentricities out of their sound for this album, but it was still full of left-wing politics dealing with the environment and struggles faced by Australia's native Aborigines (the album came out two years after the band toured indigenous settlements with the Australian Warumpi Band). The sleeve, of course, continues the mix of deeper meanings with palatable presentation. At first glance, it looks like a rather paradisiacal hut in the sand. Look again, however, and it is much bleaker; a rudimentary house stuck in the middle of nowhere makes sure that the buyer is under no illusions as to how seriously the music should be taken.

RECORD LABELS

CBS, Columbia

RELEASE DATES

Aus: August 1987, UK: April 1988, US: February 1988

SONGWRITERS

Peter Garrett, Peter Gifford, Bones Hillman, Rob Hirst, Andrew James, Jim Moginie, Martin Rotsey

MIDNIGHT OIL

Diesel and Dust

R.E.M.

DOCUMENT (1987)

The artwork for **Document** is itself something of a document of ideas, and certainly a collection of self-references from one of America's premier alt-rock bands. 'R.E.M. No. 5' is a more formal way of referring to what is the band's fifth LP, but it was also a rejected title for the album. The broken image, with a book spine-like design separating the left-hand quarter of the sleeve, looks like a grainy, messy collection of paraphernalia from a surveillance file. The documenting imagery is furthered by the image of Michael Stipe taking a photo of the photographer taking a photo of him. In keeping with all the self-references, the original sleeve design featured a 'File Under Fire' message, which is apt given how the album is generally seen as R.E.M.'s most fiery to date. It also references the 'File Under Water' message on their second album, **Reckoning** (1984), and the 'File Under Grain' that would appear on their **Eponymous** compilation album the following year. Other working titles for the album included *Mr Evil Breakfast*, *Table Of Content*, *Skin Up With R.E.M.* and *Last Train To Disneyland*. As messy as the sleeve may seem, it is evidence of a band working with more ideas than they could ever have the opportunity to use.

RECORD LABELS

IRS

RELEASE DATES

October 1987 (US: September 1987)

SONGWRITERS

Principal songwriters: Bill Berry, Peter Buck, Mike Mills, Michael Stipe

Secondary songwriters: B.C. Gilbert/Robert Gotobed/Graham Lewis/Colin Newman

R.E.M. NO.

5

DOCUMENT

GEORGE MICHAEL

FAITH (1987)

After announcing Wham!'s break-up in 1985, George Michael knew that his solo career had to start with a bang if he were to maintain the large fanbase he had accrued with Wham!'s mega stardom. He also needed to reinvent himself, marking a clear line of division between his new music and his old Wham! work. After spending the best part of 1987 writing the **Faith** album, Michael stepped out with a look that redefined him as a serious songwriter and pop artist, on the level of, say, Madonna, rather than simple pin-up fare as he was on Wham!'s **Fantastic** and **Make It Big** sleeves. The 80s saw a resurgence in interest in the Christian faith, and from its title down to the large cross earring hanging from Michael's left ear, and a selection of symbols in the bottom left-hand corner of the sleeve, including the Christian cross, Jewish Star of David and a heart with an arrow shot through it, Michael was presenting himself as a more adult entertainer, now bearded and thoughtful. The seductive pose ensured that he kept his teenage fanbase as well. It is the sort of image revisionism that other ex-boyband members would pick up on across the decades, most notably Robbie Williams.

RECORD LABELS

Epic, Columbia

RELEASE DATES

November 1987

SONGWRITERS

Principal songwriter: George Michael

Secondary songwriter: David Austin

SONIC YOUTH

DAYDREAM NATION (1988)

Five albums in, Sonic Youth found a new songwriting talent with 1987's **Sister**. For the following year's **Daydream Nation**, they moved to marry their songwriting with their love for noise rock and experimentation, resulting in a more subtly textured record than any they had recorded up to that point. It is a slightly transitional phase on the way to their more wholly accessible albums **Goo** (1990) and **Dirty** (1992), and the detailed sleeve does something to consolidate this. Amazingly, it is not a photo but a painting by Gerhard Richter, considered to be one of the most important post-war German artists. He intended for much of his finished work to show the reality of painting as a process. A photorealist, his paintings are based on photographic projections on to his canvases, over which he paints their exact form. Adding a soft-brush blur on the paintings, he not only gives his work that soft photographic quality, but also makes sure there is no mistaking that it was created by a human, free to do whatever they wanted. Sonic Youth no doubt found the deliberate fuzzing of edges an appealing aspect of Richter's work.

RECORD LABELS

Blast First, Torso

RELEASE DATES

October 1988

SONGWRITERS

Kim Gordon, Thurston Moore, Lee Ranaldo, Steve Shelley

U2

RATTLE AND HUM (1988)

For many critics, U2 stumbled when they released **Rattle And Hum**, something of a grab-bag of live and studio recordings, original and cover material, taken largely from when the band were touring with **The Joshua Tree**. The sleeve, however, captured the essence of the album and U2 at that moment. **Rattle And Hum** also became a live tour 'rockumentary' and the sleeve has a decidedly 'film noir' feel to it, which worked wonders with tying the album into the film release. The real beauty of it, however, is that it is the first U2 album artwork that presents them stripped-down and as they are: no large concepts, just a band playing its music and returning to its songwriting influences. The album was 'conceived as a scrapbook, a memento of that time spent in America on the *Joshua Tree* tour', Edge has recalled (they even cover the 'Star Spangled Banner' on it); the sleeve gives all the promise of live rock'n'roll energy, the band at their least conceptualized (contrast this with the Zoo TV Tour) and actually serves to make the group seem more tangible, despite being well on their way to defining stadium-rock stardom.

RECORD LABELS

Island

RELEASE DATES

October 1988

SONGWRITERS

Principal songwriters: Bono, Adam Clayton, Edge, Larry Mullen Jr.

Secondary songwriters: Bob Dylan, Francis Scott Key/John Stafford Smith, John Lennon/Paul McCartney, Macie Mabins/Bobby Robinson

U2

RATTLE AND HUM

N.W.A

"STRAIGHT OUTTA COMPTON" (1989)

"Straight Outta Compton" is one of *the* most influential albums of all time. Almost 20 years on, mainstream hip-hop artists are still copying the N.W.A look and lyrics. With its violent lyrics, sexism and unashamed graphic imagery, **"Straight Outta Compton"** created as much interest via its controversy as it did its actual content. But, like the punks a decade before, N.W.A (Niggaz With Attitude) were just singing about the street life they knew and it connected with the youth, creating another music explosion that sent the establishment running. Like the Ramones' presentation of themselves on the **Ramones** sleeve (see page 153), N.W.A rock up in their street clothes, only they turn the notch up a little. The album is full of gang violence and, instead of merely lining up and looking nasty as the Ramones did, here the idea is that you have already been victim of an N.W.A beating – and you may not even have heard the album yet. That Eazy-E is pointing a gun at you only makes it more obvious: N.W.A aren't to be messed with. On to Jay-Z's gangster imagery and Wu-Tang Clan's hip-hop posse presentation, N.W.A created one of the longest-standing looks in music history with this sleeve.

RECORD LABELS

Fourth & Broadway, Ruthless

RELEASE DATES

August 1989

SONGWRITERS

Dr Dre, Eazy-E, Ice Cube, MC Ren

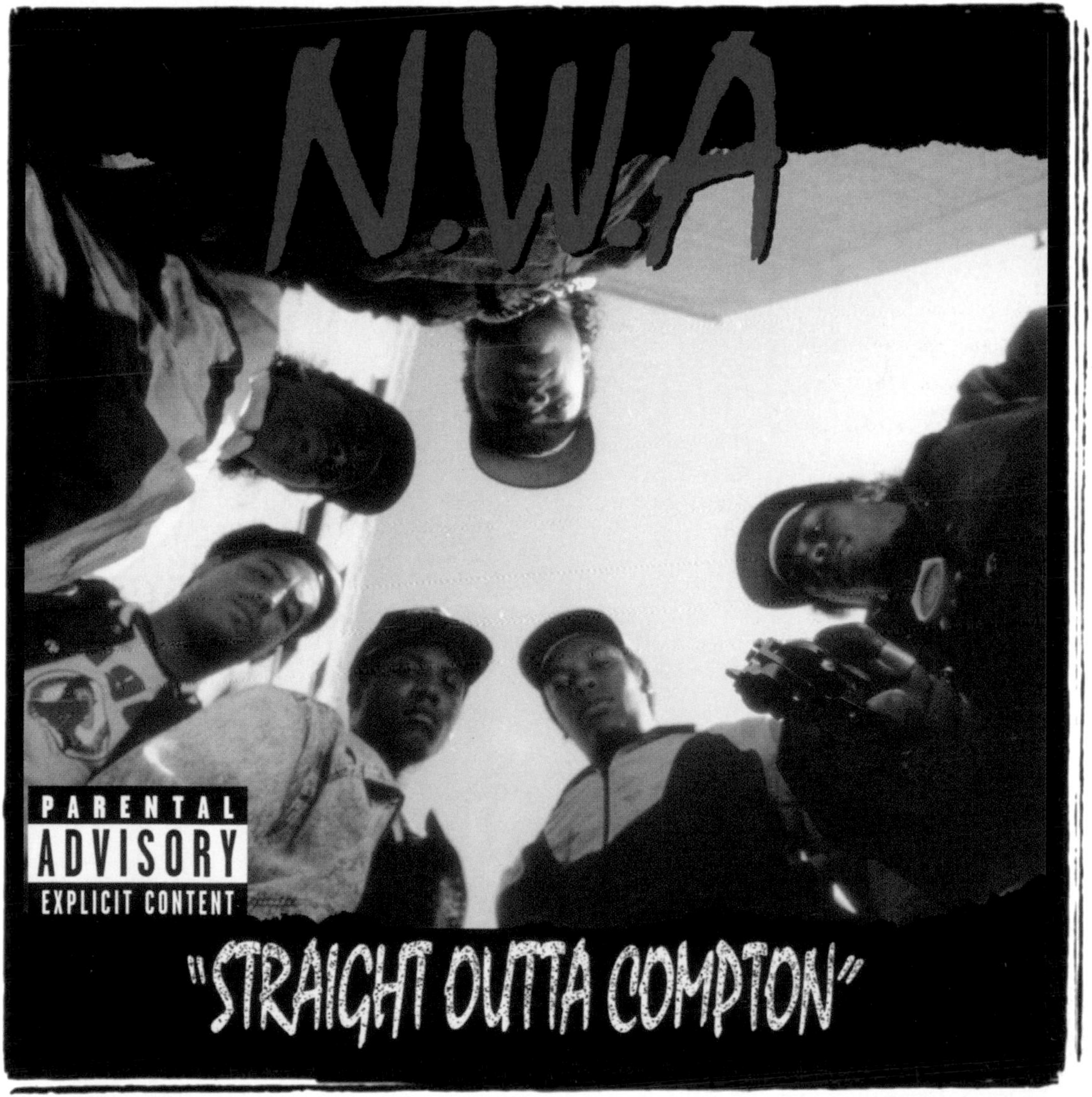
N.W.A
PARENTAL
ADVISORY
EXPLICIT CONTENT
"STRAIGHT OUTTA COMPTON"

ALBUM COVERS

RIFF
ORY

NINETIES

289

PUBLIC ENEMY

FEAR OF A BLACK PLANET (1990)

From their 1987 debut, **Yo! Bum Rush The Show**, through 1988's **It Takes A Nation Of Millions To Hold Us Back** and on to their third album, **Fear Of A Black Planet**, Public Enemy's sleeves played on increasingly grand images of military invasion. **Bum Rush...** showed the PE posse ready to do just that, **Nation...** put PE mainmen Chuck D and Flavor Flav in a jail cell that surely wasn't going to hold them, and with **Fear Of A Black Planet**, the message is writ large: the Public Enemy invasion has begun, and the world isn't ready for it. The black planet with the stylish Public Enemy logo encroaches on Earth, promising a stratospheric collision on a much more violent scale than anything George Clinton came up with. Clinton was talking about a mothership *connection*, whereas Public Enemy were looking to raze and burn. The rear sleeve saw the group in a boardroom, dressed in military clothing and plotting their attack. Given the 'Go back to Africa' samples on the album and tracks such as 'Welcome To The Terrordome', it is clear that PE knew the best form of defence is attack.

RECORD LABELS

Def Jam

RELEASE DATES

April 1990

SONGWRITERS

Principal songwriters: Chuck D, Eric 'Vietnam' Sadler, Keith Shocklee

Secondary songwriters: Flavor Flav, A. Hardy, D Jackson, Nile Rodgers, Hank Shocklee

INCLUDES
THE HIT SINGLES
'WELCOME TO
THE TERRORDOME'
& '911 IS A JOKE'
4662811
FEAR
OF A
BLACK
PLANET
UNTERATTACK ON WORLD SUPREMACY...THE COUNTERATTACK ON WORLD SUPREMACY...THE COUNTERATTACK ON WORLD SUPREMACY...THE COUNTERATTACK ON

LL COOL J

MAMA SAID KNOCK YOU OUT (1990)

With his previous album, **Walking With A Panther** (1989), LL Cool J went some way to chasing the popular hip-hop trends, losing some of the purists that had followed him since 1985's hard-edged **Radio**. The main disappointment was LL's apparent wish to throw himself towards commerciality, not least in introducing rap ballads into his repertoire, in order to appeal to a female R&B audience. By the time of **Mama Said Knock You Out**, LL's authenticity was being questioned in some quarters, though the album largely saw him return to his street roots. The hard-hitting title track, a reference to LL's ongoing feud with Kool Moe Dee (who claimed that LL stole his rapping style), was written after his grandmother told him to knock out the critics who thought his career was in decline. The sleeve somewhat mirrors Isaac Hayes' **Hot Buttered Soul** (see page 75), but with a much grittier edge, thanks to its dark, monochrome photo and the huge 'LL Cool J' knuckle-duster, while LL, his eyes hidden, cuts an imposing figure. But that has been LL's talent all along; not many hip hoppers have had a career as long as he, and reinventing himself just when he needs it most marks him out as a survivor.

RECORD LABELS

Def Jam

RELEASE DATES

September 1990

SONGWRITERS

Principal songwriters: James Todd Smith, Marlon Williams

Secondary songwriters: Brian Latture/Dwyane 'Muffla' Simon

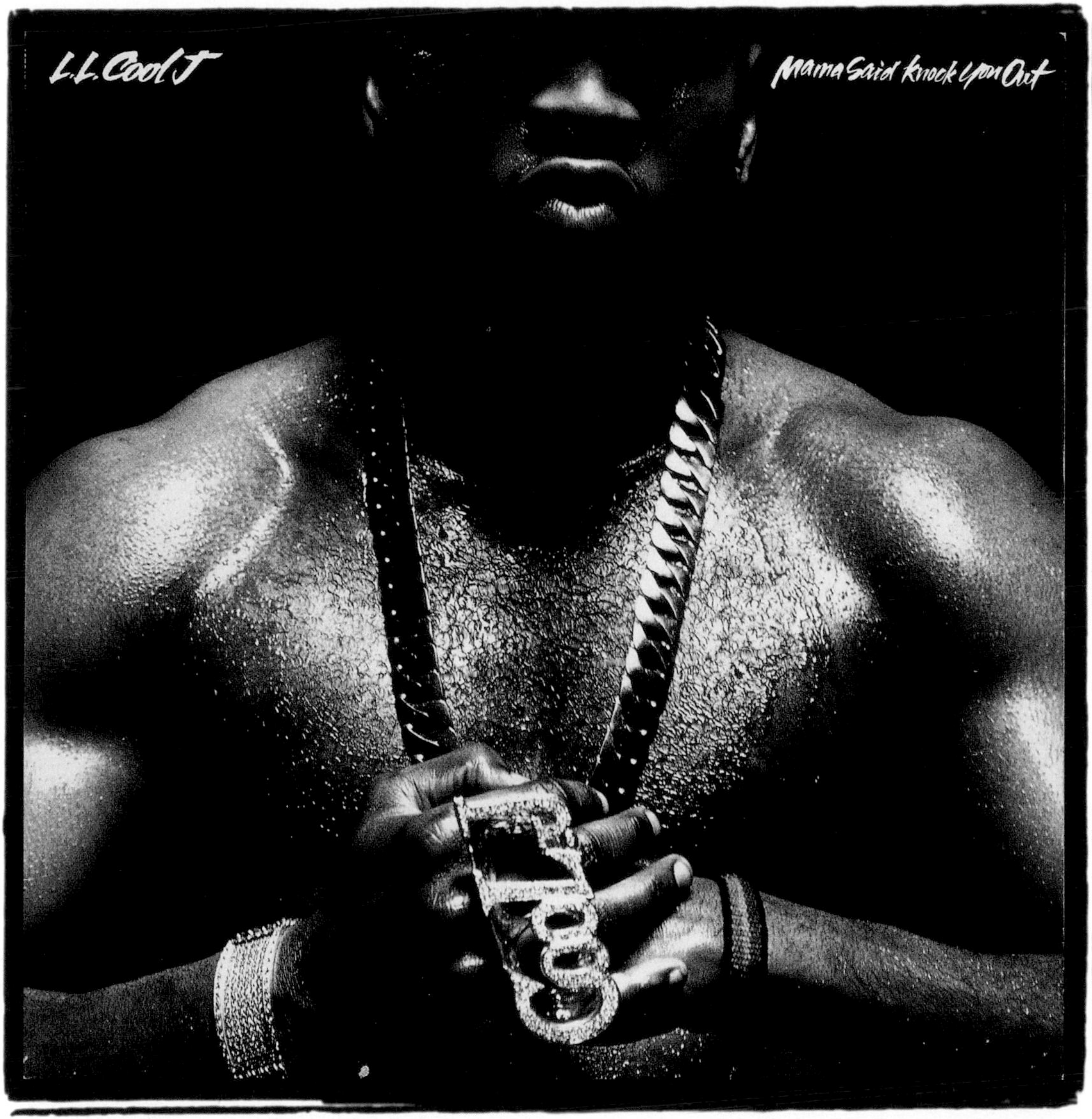
L.L. Cool J
Mama Said Knock You Out

NIRVANA

NEVERMIND (1991)

Kurt Cobain's sense of humour was certainly intact when he created the sleeve for **Nevermind**, the album which would make his band global superstars. As a commentary on record companies and capitalism in general, the dollar bill on a fish hook luring the innocent baby in at such a young age speaks volumes about Cobain's dissatisfaction with the record industry. He got the idea after watching a documentary on water births with Nirvana's drummer Dave Grohl, though the money and hook was initially a joke. Nirvana's record label, Geffen, were worried that the baby's exposed penis would offend the public, so mocked up a second cover that didn't show it. Cobain refused to use it, however, demanding that if a compromise were to be made, it had to be a sticker over the sleeve which read, 'If you are offended by this, you must be a closet paedophile.' The baby on the sleeve was three-month-old Spencer Elden, son of Geffen art director Robert Fisher's friend Rick Elden.

RECORD LABELS

Geffen

RELEASE DATES

September 1991

SONGWRITERS

Principal songwriter: Kurt Cobain

Secondary songwriters: Dave Grohl, Krist Novoselic

PRIMAL SCREAM

SCREAMADELICA (1991)

Individuality had become so key to a band's image in the 90s that no one sleeve really identified the decade in the same way that Cyndi Lauper's **She's So Unusual** (see page 253) would define young girls' fashion in the 80s. Many iconic sleeves defined a movement, though, and Primal Scream's **Screamadelica** perfectly captured the early 90s acid-house *zeitgeist*. Like Nirvana's **Nevermind** did for grunge across the Atlantic, **Screamadelica** brought dance music into the mainstream in the UK, and the bright, primary colours of its sleeve mirrored the bright, positive mentality of early 90s dance culture, and the mind-blowing effects of drug consumption that club-goers would usually experience. The dilated pupils illustrated the heavy Ecstasy intake that would pretty much fuel the underground dance culture up until the tragic death of eighteen-year-old Leah Betts, who fell into a coma after taking Ecstasy on her eighteenth birthday. For the time being, though, **Screamadelica** ushered in a new Primal Scream at odds with the indie rock of their previous two albums. From this point the Scream would be hailed as some sort of travelling band of drug visionaries, an image which they seem all too keen to perpetuate.

RECORD LABELS

Creation, Sire

RELEASE DATES

September 1991

SONGWRITERS

Principal songwriters: Bobbie Gillespie, Andrew Innes, Robert Young

Secondary songwriters: Roky Erickson/Tommy Hall

RED HOT CHILI PEPPERS

BLOOD SUGAR SEX MAGIK (1991)

The 90s saw a resurgence of West-Coast American rock, largely thanks to the popularity of Red Hot Chili Peppers among surfers (the band would prove hugely popular in the British West Country, where surfing was rife in towns such as Newquay). Part of the 'surfer dude' imagery incorporated elaborate tattoos, and **Blood Sugar Sex Magik** saw the band revelling in the art, while creating their most memorable and iconic sleeve (not too difficult, given that their other artworks are largely forgettable) for their most essential album. The interlocking tongues design was drawn by Henk Schiffmacher, a tattoo artist who had designed most of the band members' own tattoos, while the cover photo was taken by Gus Van Zant, director of the 'Under The Bridge' promo video. The album's booklet largely consisted of a collage of each band member's many and varied tattoos, making the Chilis partially responsible for the 90s craze for having tattoos based around Chinese symbols and other strange imagery that wasn't usually found on your everyday heavy metallist's bicep.

RECORD LABELS

Warner Bros.

RELEASE DATES

September 1991

SONGWRITERS

Principal songwriters: Flea, John Frusciante, Anthony Kliedis, Chas Smith

Secondary songwriter: Robert Johnson

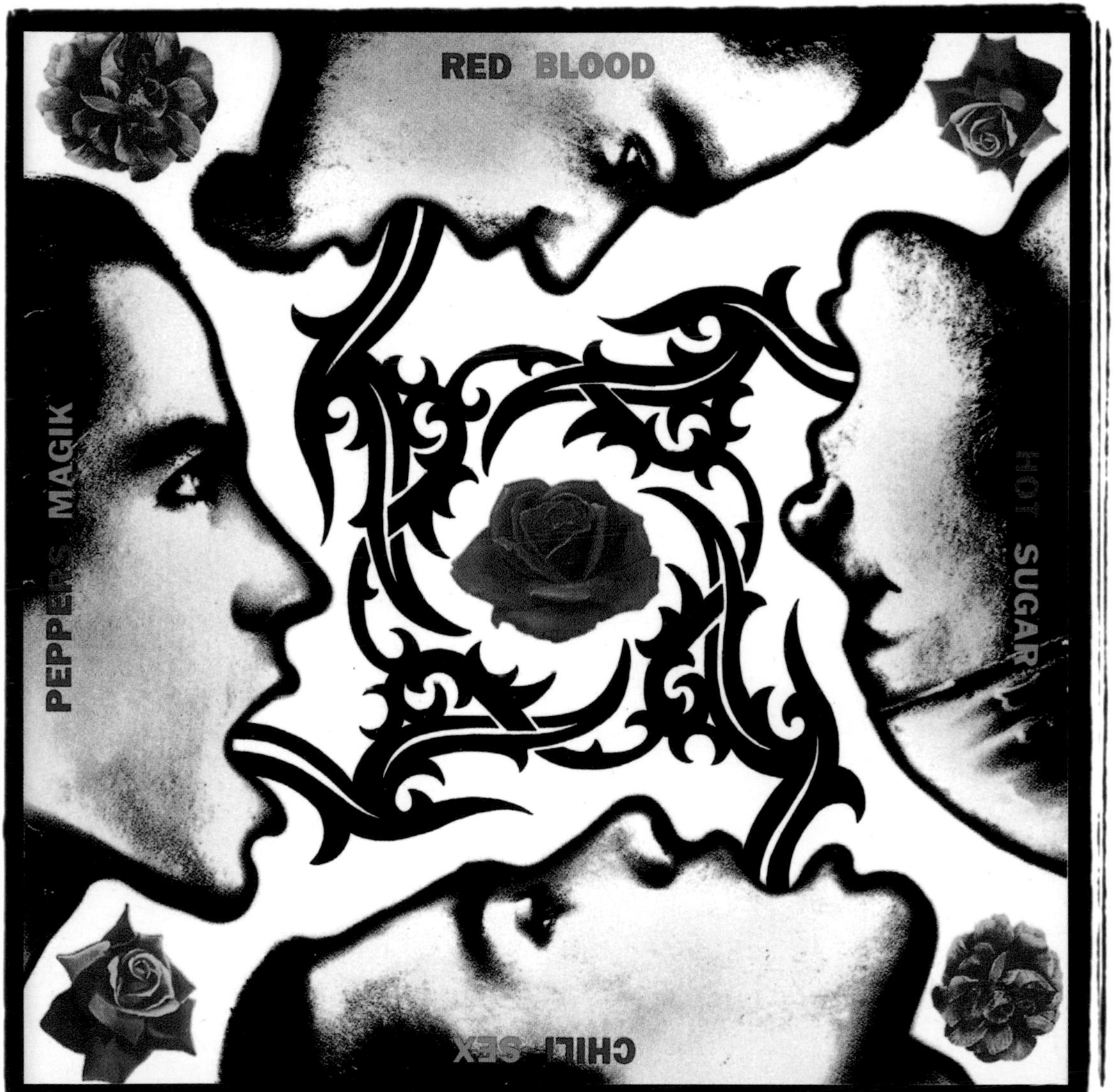

RED BLOOD
HOT SUGAR
CHILI SEX
PEPPERS MAGIK

U2

ACHTUNG BABY (1991)

After a long line of po-faced, black-and-white album sleeves attesting to U2's solemnity and the seriousness with which they took their mission, **Achtung Baby** made you pay attention once again, in case you were beginning to take their more statement-making sleeves for granted. The album itself saw the band embracing electronica and dance idioms for the first time, while also working heavily with guitar effects. In short, they took cues from the Primal Screams, Happy Mondays and My Bloody Valentines of the day. As such, the brightly coloured collage for **Achtung Baby** reveals something of a more populist U2, keen to grab at the UK charts in the new decade. It also goes some way to suggesting that the music is not full of political rhetoric, but focused more on personal lyrical content. All round it presents a much more experimental U2 than previously seen. The collage artwork mirrors the more adventurous take-from-anywhere overtones of the music, and also goes some way towards ushering in the scrapbook-like openness of Bono and Edge's lyrics at the time.

RECORD LABELS

Island

RELEASE DATES

November 1991

SONGWRITERS

Bono, Adam Clayton, Edge, Larry Mullen Jr.

SPIRITUALIZED

LAZER GUIDED MELODIES (1992)

Spiritualized essentially comprised Jason Pierce, one of the founding members of Spacemen 3, and whichever ensemble of musicians he chose to work with at any given time. Spacemen 3 worked in a post-punk, drone-rock field, with the motto 'Taking drugs to make music to take drugs to', and a minimalist sound that involved avoiding chords in their arrangements. In many ways Spiritualized became much more expansive, with Pierce later incorporating gospel into his sound. For **Lazer Guided Melodies**, however, he expanded upon his contributions to Spacemen 3, with four extended suites that built upon his skills as a songsmith. The sleeve mirrors the space-age dance music vibe, with loose, liquid bodies shimmering as much as the music does. It is still music to take drugs to, however, and with the right combination, and if in the right frame of mind, **Lazer Guided Melodies** could turn you into as loose-limbed a character as appears on the front. It is a dangerous line to tread, and the devil seemingly preying on the voluptuous woman suggests that this isn't music to go into lightly. In just one year, the sort of drug high afforded by **Screamadelica** had become a more negative beast entirely to some people.

RECORD LABELS

Dedicated

RELEASE DATES

April 1992

SONGWRITERS

Principal songwriter: Jason Pierce

Secondary songwriters: J.J. Cale, Mark Refoy

RAGE AGAINST THE MACHINE

RAGE AGAINST THE MACHINE (1992)

From their name down to the once-heard-never-forgotten 'Fuck you, I won't do what you tell me' mantra, everything about Rage Against The Machine's debut spelt defiance and revolution. Even their mix of rap, hip hop and funk was something of a trend-bucker, and was hugely influential upon later – though much more watered down – groups such as Limp Bizkit. Not known for their subtlety, the sleeve for their debut was a black-and-white photo of Thích Quảng Đức the Vietnamese Mahayana Buddhist monk who burned himself to death in Saigon, protesting against the Buddhist oppression by President Ngô Đình Diệm's South Vietnam administration. Rage Against The Machine so strongly identified themselves with anyone that questioned any authority that they also thanked Black Panther founder Huey P. Newton in the sleevenotes for their debut, along with Bobby Sands, a Provisional Irish Republican Army volunteer who was imprisoned in HM Prison Maze for possession of firearms, but who died on hunger strike. Later Rage sleeves would become less and less visually polemical, but for a debut release, this is as arresting as they come.

RECORD LABELS

Epic

RELEASE DATES

February 1993 (US: November 1992)

SONGWRITERS

Tim Commerford, Zack de la Rocha, Tom Morello, Brad Wilk

AEROSMITH

GET A GRIP (1993)

A long line of wry, title-referencing sleeves have seen Aerosmith albums come with artwork of two trucks seemingly having sex (**Get A Grip**'s predecessor, **Pump** [1989]), the caricature-style pen drawing of **Draw The Line** (1977), and the what-goes-on-at-night **Toys In The Attic** (1975) sleeve. **Get A Grip** is pretty much inviting despite itself. The 'Aerosmith'-branded cow, with udders close-up, isn't so appealing, but then the ring-pierced udder will set a certain type of mind racing. Heavy-metal circles are full of pranksters and many nipple-pierced men will no doubt have fallen foul of the old pull-the-ring-really-hard gag, which results in a lot of pain and blood loss for the victim. It is probably not too wild a guess to assume that Aerosmith, notorious for their hard-partying lifestyle, have seen this happen once or twice, and here drawn up a sleeve so enticing for the adolescent-minded that getting a grip on the udders themselves is probably not even considered an option. As a nice touch, promo copies of the CD were sent out in synthetic cowhide sleeves, with each one allegedly featuring a different pattern.

RECORD LABELS

Geffen

RELEASE DATES

April 1993

SONGWRITERS

Principal songwriters: Joe Perry, Steven Tyler

Secondary songwriters: Jack Blades, Desmond Child, Mark Hudson, Lenny Kravitz, Taylor Rhodes, Tommy Shaw, Richard Supa, Jim Vallance

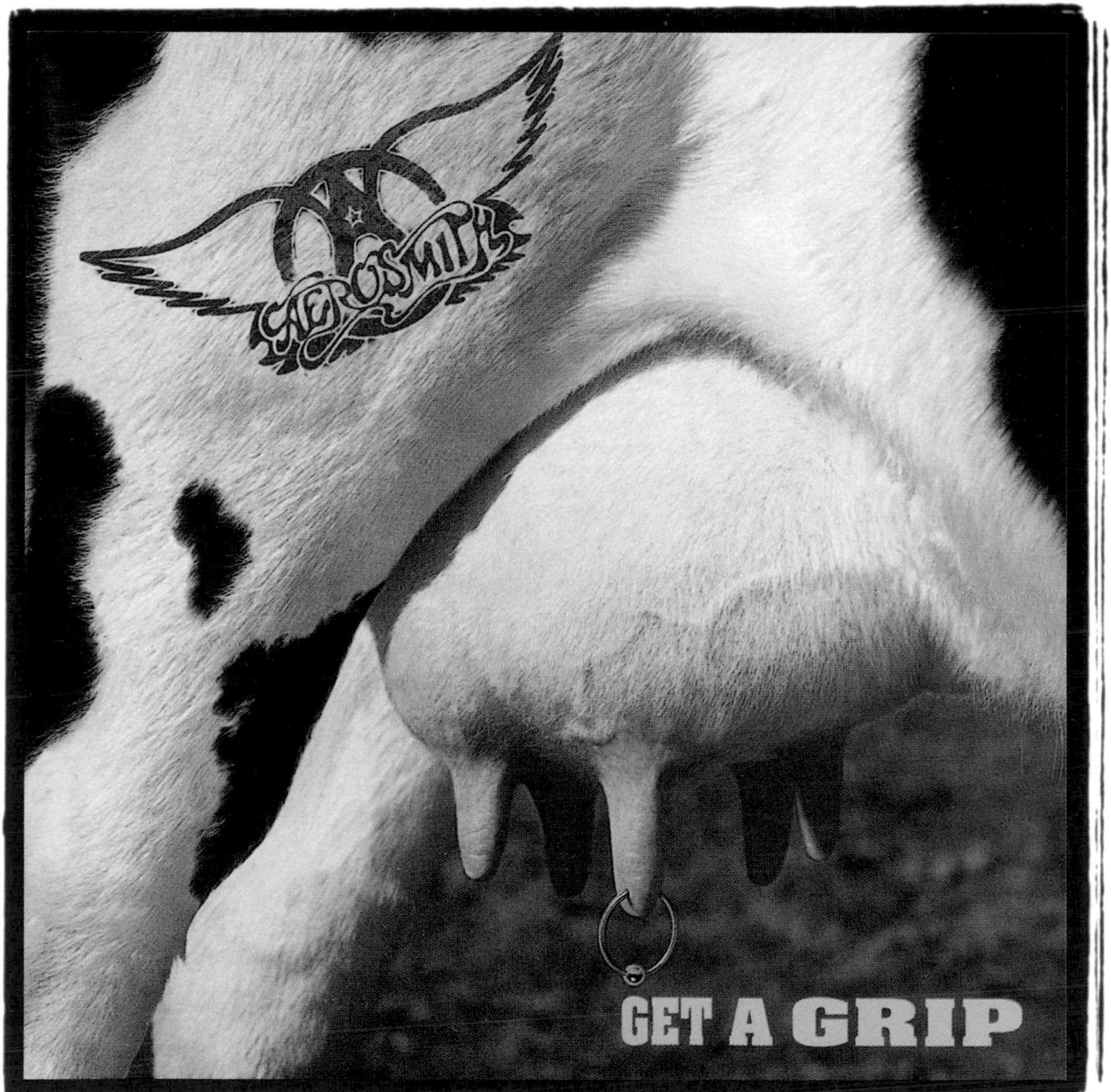
AEROSMITH
GET A GRIP

BJÖRK

DEBUT (1993)

Debut wasn't exactly Björk's debut. One solo album had preceded it in her homeland of Iceland but, internationally speaking, **Debut** was the first the majority of the UK had heard of the ex-Sugarcubes singer. Free from her Sugarcubes duties (and their sleeves, which simply presented the group's name on brightly coloured, Pop Art-style backgrounds), the **Debut** sleeve presents a much more unassuming singer than is evidenced in the album's music. The black-and-white, hands-together pose is almost thankful, almost hopefully yearning for you to like her 'debut' offering. It also demands that Björk be taken as a serious songwriter. Given the dance pop nature of much of the music, it might not have happened if the artwork were more frivolous. In hindsight it is easy to see **Debut** as a real jumping-off point, and not in fact the 'real' Björk at all. It is perhaps an easy introduction that allowed her, once accepted, to further out-weird herself on later sleeves, which see her in all manner of increasingly bizarre costumes, culminating most recently in the Easter egg/chicken suit on **Volta** (2007).

RECORD LABELS

One Little Indian, Elektra

RELEASE DATES

July 1993

SONGWRITERS

Principal songwriter: Björk

Secondary songwriters: Nellee Hooper, Johnny Burke/James Van Heusen

SNOOP DOGGY DOGG

DOGGYSTYLE (1993)

It is amazing to think that Snoop Dogg has become something of an elder statesman of hip hop. Although not exactly clean-cut today, his image is much more palatable to the public now than it was in the early 90s days of gangsta rap; Snoop Dogg was arrested on murder charges the same year that **Doggystyle** came out. **Doggystyle**'s influences are literally worn on its sleeve, while the title also references the sexual act that looks imminent. Along with the likes of Digital Underground, who began looking to the P-Funk catalogue in the late 80s, producer Dr Dre (who brought Snoop to the world's attention on his 1992 album **The Chronic**) was at the vanguard of West-Coast producers who sampled George Clinton's back catalogue. The sleeve is an early 90s take on Pedro Bell's Funkadelic artwork, while the conversation between the three dogs on the wall is made up of three direct quotes from Clinton's biggest solo hit, 1982's 'Atomic Dog' (Snoop Dogg's 'Who Am I (What's My Name)?' is largely built upon 'Atomic Dog', along with one Funkadelic and two Parliament tracks). Snoop's music would come to typify the strand of West-Coast hip hop known as G(angsta)-Funk, again a name chosen thanks to its overt homage to Clinton's work.

RECORD LABELS

Death Row-East West, Death Row-Interscope

RELEASE DATES

December 1993 (US: November 1993)

SONGWRITERS

Principal songwriters: Dr Dre/Snoop Dogg

Secondary songwriters: Delmar Arnaud/R. Brown/Bootsy Collins/Parker/Andre Young, Larry Blackmon, Harry Wayne 'K.C.' Casey/Richard Finch, George Clinton/Gary Shider/David Spradley, Dat Nigga Daz, Douglas Davis/Hachidai Nakamura/ Ei Rokusuke/Ricky Walters, Richard 'Dimples' Fields, Warren G/Kurupt

UNDERWORLD

DUBNOBASSWITHMYHEADMAN (1993)

Along with the Orb, Underworld was one of the main catalysts for dance music's mainstream popularity in the 90s. Mainmen Karl Hyde and Rick Smith are also founding members of the graphic design company Tomato, which, since its early 90s inception, has helmed many art installations and designed artwork for television and print advertising, along with corporate logos. Tomato also designed every Underworld sleeve from **dubnobasswithmyheadman** onwards. Compare this to the unremarkable sleeves for 1988's **Underneath The Radar** and 1989's **Change The Weather**, and you will see how important that is. The heavily stylized and textured sleeve, contrasting the regimented with the organic, is a perfect visual accompaniment to the collision of musical styles and sound effects that the album offers. **dubnobasswithmyheadman** presents itself as heavily detailed music much more than the sum of its parts. Many less eclectic releases would also use such intricate graphic design in the 90s as a way of illustrating just how many genres the groups span (see the sleeves of Primal Scream's **Vanishing Point** and its remix album, **Echodek** [both 1997], the latter of which plays heavily on the black and white of **dubnobasswithmyheadman**), while this sleeve is something of a template for bands that would release dance music created with a human heartbeat.

RECORD LABELS

Junior Boys Own

RELEASE DATES

December 1993

SONGWRITERS

Principal songwriters: Karl Hyde/Rick Smith

Secondary songwriter: Darren Emerson

dubnobasswithmyheadman
underworld
dubnobasswithmyheadman
jbolp1
underworld
naked
and angel
for a perfectstar
she was beautiful god
she was beautiful god
she was beautiful god
and in walks this girl
real lolita
just groan chemical
derail me every time.
bass m.e.
dubnobasswithmyheadman
underworld

GREEN DAY

DOOKIE (1994)

Though Nirvana kicked down the door, Green Day are largely responsible for the 90s punk-pop revival. Now hailed as serious punk agitators commenting on the American condition, back in 1994 Green Day were largely still just kids who found amusement in writing songs about masturbation. The title was a reference to the band suffering from diarrhoea whenever they went out on tour (something they called 'liquid dookie'). The intricate sleeve design features bombs being dropped on the East Bay, California punk-pop scene in which the band grew up, with Green Day exploding out of the wreckage. It was designed by local artist Richie Bucher to 'represent the East Bay and where we come from', Green Day mainman Billie Joe Armstrong has said. 'There's pieces of us buried in the album cover. There's one guy with his camera up in the air taking pictures with a beard. He took pictures of bands every weekend at Gilman's [a Berkeley music foundation] … [AC/DC's] Angus Young is in there somewhere too. The graffiti reading "Twisted Dog Sisters" refers to these two girls from Berkeley'. It is a 90s punk-pop cross between a Robert Crumb illustration and **Sgt. Pepper's Lonely Hearts Club Band** (see pages 61, 49) for a local music scene.

RECORD LABELS

Reprise

RELEASE DATES

February 1994

SONGWRITERS

Billie Joe Armstrong/Mike Dirnt/Tré Cool

OASIS

DEFINITELY MAYBE (1994)

Like many great British rock bands (fellow Mancunians the Smiths included), Oasis landed fully formed, in sound and image. By the end of 1994 Ben Sherman shirts would be back in fashion, and Britain's obsession with the Beatles would return in full force. To the normal eye, this just looks like an arty shot of a band relaxing in a front room, with a cool logo that is instantly recognizable. When people realized that Oasis were Beatles obsessives, however, it became *de rigeur* to read meanings into their album sleeves. What did they get from **Definitely Maybe**? Not a whole lot, though given the group's later excesses, red wine is a rather more civilized drink than might have been expected. The George Best photo is a tribute to one of Manchester United's most celebrated players (who also appeared on the sleeve for the Wedding Present's **George Best** [1987]), while the Burt Bacharach poster references one of songwriter Noel Gallagher's other main inspirations. For his part, singer Liam Gallagher lays out in a pose similar to John Lennon in a famous mid-60s Beatles poster (which went back into heavy print rotation during the mid-90s). In all then, it defines Oasis: a group very much made up of their influences.

RECORD LABELS

Creation, Epic

RELEASE DATES

August 1994 (US: January 1995)

SONGWRITERS

Noel Gallagher

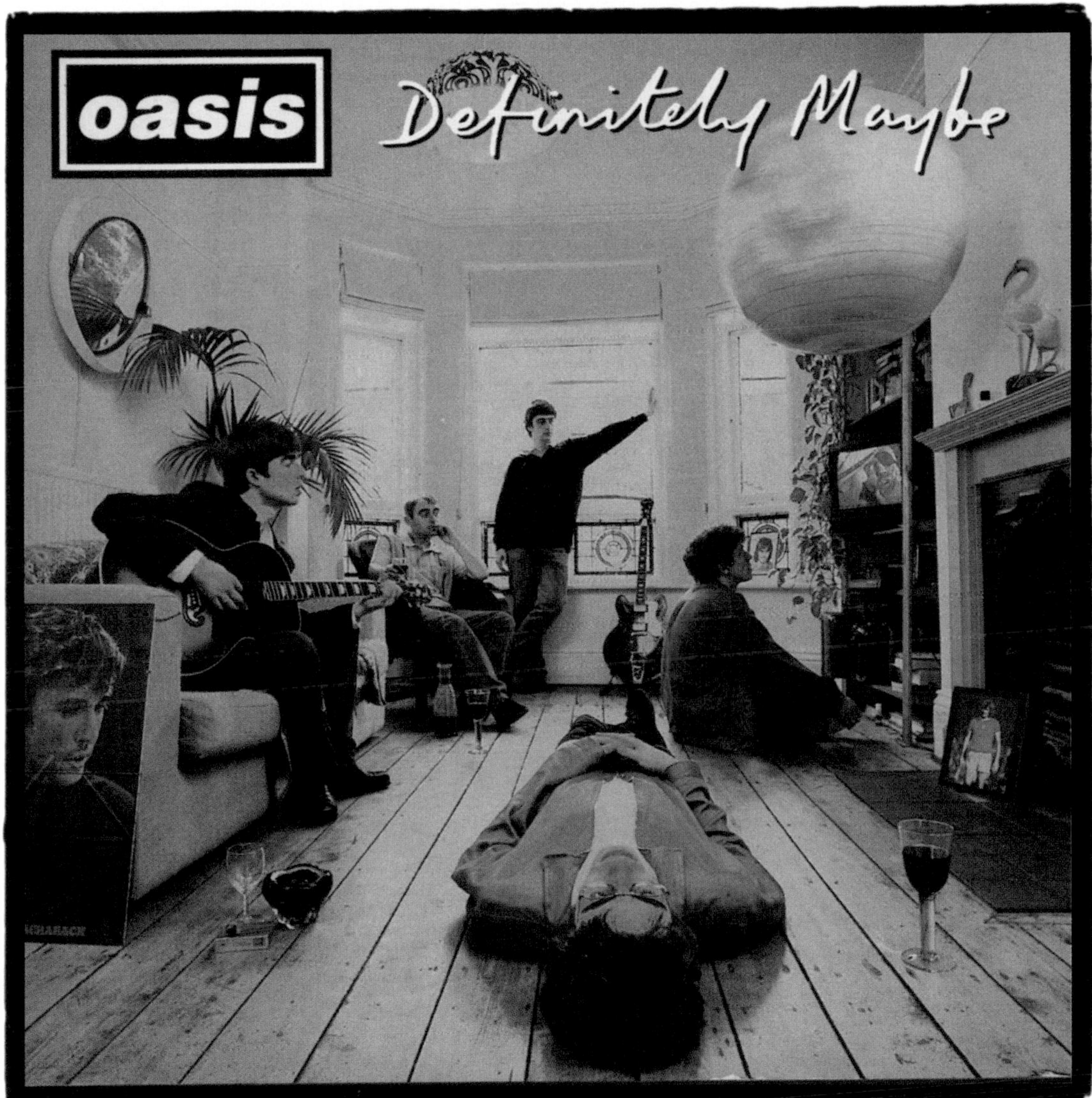
oasis
Definitely Maybe

BLACK CROWES

AMORICA (1994)

For the album that broke them, Black Crowes made sure they got attention. Three years before the advertising campaign for *The People Vs. Larry Flynt* biopic would more closely bring pornography and American iconography together, the Crowes' took an image from a 1976 issue of Flynt's porno magazine *Hustler*, which celebrated the United States Bicentennial in its own way. Seems innocent enough, but the fact that the model's pubic hair was on show still caused outrage in the early 90s, and the Crowes' label, American Recordings, put the album out with an edited version of the sleeve. It was an image so at odds with their previous of-the-earth sleeves of **Shake Your Money Maker** (1991) and **The Southern Harmony And Musical Companion** (1992) that it could not help but gain notice. As for **Amorica**, that was the name given to the Brittany Peninsula part of ancient Gaul, where the comic-book character Asterix lived. Were they already so sure that the sleeve would court controversy that they decided to design one pointing out just how ancient America's traditions were?

RECORD LABELS

American

RELEASE DATES

November 1994

SONGWRITERS

Chris Robinson/Rich Robinson

JOHNNY CASH

AMERICAN RECORDINGS (1994)

Throughout much of the 70s and 80s, Johnny Cash's music suffered largely from over-production, even though he had recorded reams of solo acoustic music in the mid-70s that remained unreleased at the time. As the string arrangements and country music clichés of some of his 80s work fell out of favour with the public, and even his own record label, in the 90s Cash was in danger of becoming consigned to the history books. Thank god for Rick Rubin, then, who brought Cash back to the public eye with a series of stripped-down American albums that put the focus squarely on Cash the singer and guitarist. The first, **American Recordings**, plays heavily on Cash's Man In Black image, and the seemingly continual fight between good and evil that not only featured in most of his songs, but also continued to rage on in his personal life – as symbolized by the white and black dogs at Cash's feet. In this photo he appears as judge, jury and executioner in one. The sparse, black-and-white surroundings announce that Cash is returning to his roots, while the large, bold 'CASH' hovering somewhere in the sky is all that needs to be said, really.

RECORD LABELS

Rhino, American

RELEASE DATES

1994

SONGWRITERS

Principal songwriter: Johnny Cash

Secondary songwriters: Glenn Danzig, Jimmie Driftwood, Kris Kristofferson, Alan Lomax/John A. Lomax/Roy Rogers/Tim Spencer, Nick Lowe, Karl Silbersdorf/Dick Toops, Loudon Wainwright III, Tom Waits

NINETIES

OL' DIRTY BASTARD

RETURN TO THE 36 CHAMBERS (1995)

Ol' Dirty Bastard (or Russell Tyrone Jones) was always the wild card in the Wu-Tang Clan pack, a multi-project hip-hop collective that, in conception at least, were hip hop's P-Funk for the mid-90s. ODB's continued legal troubles marked him out as the member with the most media attention, thanks to a string of jail sentences, assaults and drink- and drug-related mishaps. In 1995 he was the second Wu-Tang member to release a solo album, a record which, though raw in sound, is wholly based around ODB's unpredictable stream-of-consciousness-style rapping and oddball humour. The album was recorded while Jones was still collecting welfare allowance in America, and the sleeve features his actual welfare card at the time. Proving that ODB really didn't give a hoot, his erratic behaviour led him, in 1997, to collect his latest welfare cheque while being filmed for an MTV biography. **Return To The 36 Chambers** was still in the Top 10 in the American charts, and Dirty took two of his 13 children to the office in a limousine when collecting the cheque. Captured on MTV's cameras, the whole thing was broadcast nationwide.

RECORD LABELS

Atlantic, Elektra

RELEASE DATES

March 1995

SONGWRITERS

Principal songwriters: Robert Diggs, Russell Jones

Secondary songwriters: E. Chambers, F. Cuffie, G. Grice, D. Harris, R. Jones, Raymond Reed, V. Sims, T. Starks, O. Turner

BLACK GRAPE

IT'S GREAT WHEN YOU'RE STRAIGHT ... YEAH (1995)

Shaun Ryder was probably as surprised as anyone that his post-Happy Mondays group created such a joyous, infectious debut as **It's Great When You're Straight ... Yeah**. Instead of falling into further drug abuse and ending up as a Britpop postscript (the album title seems to suggest that Ryder had found a new creativity in cleaning up his act), three years after the Mondays' decline Ryder took his multicoloured hues with him and came out with a sleeve that reflects the hedonistic party sound of the music. Never one to miss a sly joke, however, Ryder chose to do up a famous photo of international terrorist Carlos the Jackal in Pop-Art colours, suitably fitting in with the 60s-referencing Britpop styles of the time, while also making sure it wasn't as fey as some of his contemporaries' throwback imagery. Interestingly, the sleeve itself proved inspirational elsewhere. Starting in 2006, a comic-book series entitled *Phonogram* was produced in America, taking Britpop-era Britain as its main inspiration. Each issue's cover was based on a famous Britpop album sleeve, and the second *Phonogram* issue, *Can't Imagine The World Without Me*, used this sleeve.

RECORD LABELS

Radioactive

RELEASE DATES

August 1995

SONGWRITERS

Principal songwriter: Shaun Ryder

Secondary songwriters: Paul Leveridge, Stephen Lironi, Danny Saber

BLACK GRAPE
IT'S GREAT WHEN YOU'RE STRAIGHT YEAH

BECK

ODELAY (1996)

As one of the most important albums of the 90s, Beck's **Odelay** ushered in an era of wilful eclecticism in music, truly being an album with no genre boundaries. Now something of a cliché in music, at the time a record that that saw hip hop sit so comfortably with country, funk, metal and other musical experimentations (including Serge Gainsbourg influences) all on one album, was something of a revelation – especially as grunge and nu metal were riding high in the American charts. Initially, Beck intended for the album to be titled 'Aurele', after a lyric in the song 'Hotwax', which calls out the Mexican slang word, roughly translating as 'salut' or the Jewish 'l'chaim'. The sound engineer mis-heard him, however, and wrote down 'Odelay', which stuck. It was long-questioned what the cover photo was, with many thinking it an art sculpture, in keeping with the sleeve for Beck's previous major label release, **Mellow Gold** (1994). It is, in fact, a real photo, featuring a Hungarian dog known as a Komondor, known for its thick, matted, mop-like coat of hair, jumping over a hurdle in a Crufts-like competition.

RECORD LABELS

DGC

RELEASE DATES

June 1996

SONGWRITERS

Principal songwriter: Beck

Secondary songwriters: John King, Michael Simpson

PEARL JAM

NO CODE (1996)

Pearl Jam went to town with the artwork for **No Code**, their fourth album and the last one to hit No. 1 on the American *Billboard* charts. For part of the packaging they relied on Lomographic camerawork, which emphasized laid-back, almost on-a-whim photography taken on Lomographic cameras, deliberately distorting photographs with their fish-eye lenses or saturated colours. Inside each CD and LP release of the album the lyrics to each song came on the back of nine Polaroid photos. As there were actually 13 songs in total, however, and four different sets of Polaroids for each one, it took a lot of effort to get all 52 photos. The sleeve itself, made up of 144 photos, was in keeping with previous Pearl Jam albums, as the group liked to put hidden messages relating to the album's themes in their artwork. Here the sleeve folds out to reveal the eyeball-in-a-triangle logo, reflecting the meditations on spirituality and mortality that crop up throughout the album's lyrics, which saw Vedder temper the rage he unleashed on previous albums to embrace a religion of his own conception.

RECORD LABELS

Epic

RELEASE DATES

September 1996

SONGWRITERS

Principal songwriter: Eddie Vedder

Secondary songwriters: Jeff Ament, Mike McCready, Stone Gossard, Jack Irons

RADIOHEAD

OK COMPUTER (1997)

For fans of the band, Radiohead's conceptual artwork has become as renowned as its music, seeing Thom Yorke (under the pseudonym Dr Tchock) collaborate with artist Stanley Donwood on designs that each further the ideology to be found on their respective albums (a collection of their works was published in 2007 as *Dead Children Playing*). Three albums in, at the time of its release **OK Computer** saw Radiohead's most ambitious artwork to date, a mixture of computer-generated collages as designed by Yorke, and hand-drawn pieces created by Donwood. Taken separately, the collision of symbols, hidden texts and distorted, disturbing imagery says very little, but together it perfectly reflects the concerns with consumerism, globalization and paranoia in the modern world, giving the album as a whole – music and artwork – a continuity that marks it out as a whole package, as opposed to a collection of songs. In 1997 Yorke explained the **OK Computer** artwork to *Select* magazine, saying that, 'Someone's being sold something they don't really want, and someone's being friendly because they're trying to sell something. That's what it means to me. It's quite sad, and quite funny as well…. It was all the things that I hadn't said in the songs.'

RECORD LABELS

Parlophone, Capitol

RELEASE DATES

June 1997

SONGWRITERS

Colin Greenwood/Jonny Greenwood/Ed O'Brien/Phil Selway/Thom Yorke

PRODIGY

THE FAT OF THE LAND (1997)

Moving away from album sleeves more in keeping with either the sparse end of dance-music designs (1992's **Experience**) or a more ornate sleeve employed to ward off the faint-hearted and maintain an air of outsider imagery (1995's **Music For The Jilted Generation**), **The Fat Of The Land** sees the Prodigy embrace their new-found commerciality. The album brought dance music to a wider audience, thanks in no small part to the controversy kicked up with the 'Firestarter' and 'Smack My Bitch Up' singles, and even topped the *Billboard* charts in America. It also saw the group move out of their immediate dance/electronica comfort zone, working with Crispian Mills of then-popular indie group Kula Shaker, and the always entertaining king of indie hip-hop Kool Keith, while the group themselves toyed with a rock influence that appealed to indie kids, and vocalist Keith Flint adopted a more punk-derived sound and look. The sleeve comes from an all-encompassing, alt-rock angle, simply using a random image shot that has been 'artistically' distorted and left to stand on its own, with no explanation nor great deal of relevance to the album's content, other than being as stylish as the music.

RECORD LABELS

XL, Geffen

RELEASE DATES

July 1997

SONGWRITERS

Principal songwriters: Keith Flint, Liam Howlett

Secondary songwriters: Arran/Skin, Donita Sparks/James/Knight/Walsh, Kim Deal/Anne Dudley/Trevor Horn/Johnathon J. Jeczalik/ Gary Langan/Paul Morley, Kool Keith, Maxim, C. Miller, Crispian Mills, T. Randolph, M. Smith

THE FAT OF THE LAND

DAVE MATTHEWS BAND

BEFORE THESE CROWDED STREETS (1998)

It took the Dave Matthews Band four LPs to capture their well-known sound on record. Ever since 1993's **Remember Two Things** they were largely celebrated for their long, eclectic improvisations on stage, something that has never been easy to capture on record. For this album, DMB picked a sleeve that seemed to acknowledge a final coming-together of ideas on record in its generally slick presentation, but which suggested that the band hadn't wholly ironed out the creases yet, with some self-effacing coffee-stain marks. The sort of clubs the band played in would be found on streets such as the one depicted on the sleeve, and the blurry traffic image seems to reference the five years spent touring inner-city clubs, which had honed the band's collective improvisational skills. You would expect them to be saying that this street appears to be just like any other to them (as it does to the buyer), but then the mug stains almost double-up as a see-through, past the superficial busy street. It suggests that, though the streets may look the same on the outside, for the band well acquainted with such areas it is possible to see what is really going on underneath.

RECORD LABELS

RCA

RELEASE DATES

April 1998

SONGWRITERS

Principal songwriter: Dave Matthews

Secondary songwriters: Carter Beauford, Stefan Lessard, Leroi Moore, Boyd Tinsley

DAVE MATTHEWS BAND BEFORE THESE CROWDED STREETS

MASSIVE ATTACK

MEZZANINE (1998)

Given the Massive Attack 'brand image' created with the sleeves to 1991's and 1994's **Blue Lines** and **Protection** respectively, which saw each artwork almost as a Part One and Part Two based around danger signs on inflammable material, if you were told, 'Oh yeah, their next sleeve is a close-up photo of a big beetle,' you would probably think it was a joke. That said, Prodigy had made a close-up shot of a crab the sleeve for their **The Fat Of The Land** album the previous year (see page 333), so maybe something was in the air. For Massive Attack, the change in image clearly heralded a clean slate. **Mezzanine** was a brand new start announcing that the band – which had been somewhat sidelined in the public's mind in the four years since their previous album – had returned darker than ever. The stark, black-and-white beetle shot was much more in keeping with the new sound, in which rock guitars were heavily incorporated in some tracks. The fold-out triptych sleeve revealed the full extent of the insect, while said creepy-crawly also provided the artwork for single sleeves and poster campaigns linked to **Mezzanine**, helping make it clear that the album and related releases were all cut from the same cloth.

RECORD LABELS

Circa

RELEASE DATES

April 1998

SONGWRITERS

Principal songwriters: Robert '3D' del Naja, Grantley Marshall

Secondary songwriters: Michael Dempsey/John Holt/Robert Smith/Laurence Tolhurst, Elizabeth Fraser, Mort Garson/Bob Hilliard, Horace Hinds, Sara Jay, Lou Reed, Matt Schwartz, Pete Seeger, Andrew 'Mushroom' Vowles

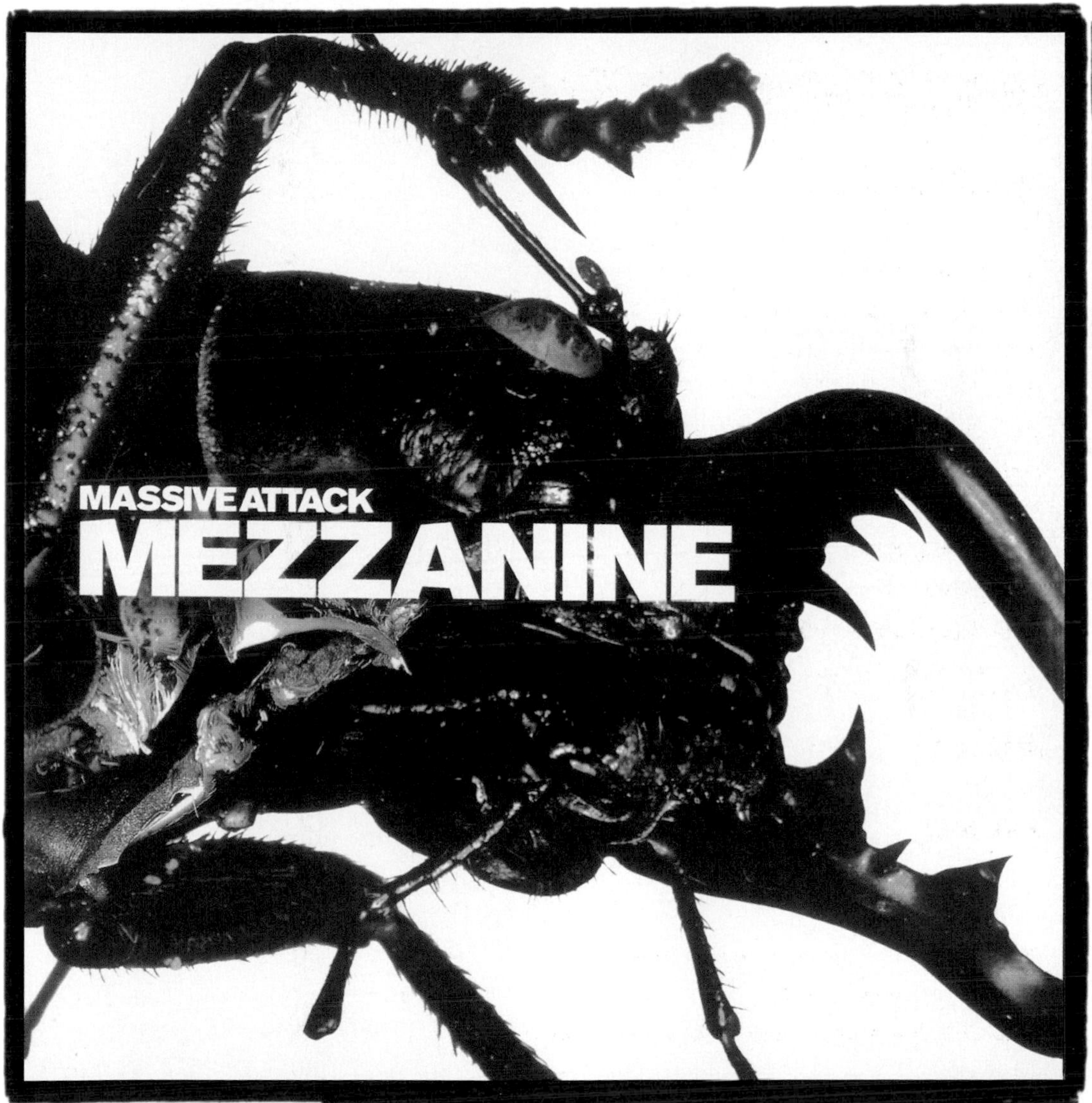
MASSIVE ATTACK
MEZZANINE

THE OFFSPRING

AMERICANA (1998)

If it had not been for Green Day's overwhelming **Dookie** success (see page 315), there would not have been a market for punk-pop bands such as the Offspring. Though they were actually Green Day's contemporaries, it took that initial breakthrough for the Offspring to even be given a chance. **Americana** proved their most popular album in the UK, largely on the back of the ubiquitous single 'Pretty Fly (For A White Guy)'. The Offspring felt they had more to say, however, and placed themselves in the long tradition of American punk-rock groups when they got Bad Religion art director Frank Kozik to direct their **Americana** sleeve. The Offspring may not have had the fuel to become serious commentators in the way that Green Day did, but their sleeve at least touches upon much of what they tried to capture: an unnerving look at suburban American life in the mid-90s, through often awkward eyes. The leg-braced child gleefully playing with a large shrimp as a tentacle reaches out for it is really only the sort of thing that punk graphic artists could come up with, and it was suitably weird enough to appeal to disaffected teens in the 90s, without them really knowing why.

RECORD LABELS

Columbia

RELEASE DATES

November 1998

SONGWRITERS

Principal songwriters: Dexter Holland, Greg K., Noodles, Ron Welty

Secondary songwriter: Morris Albert

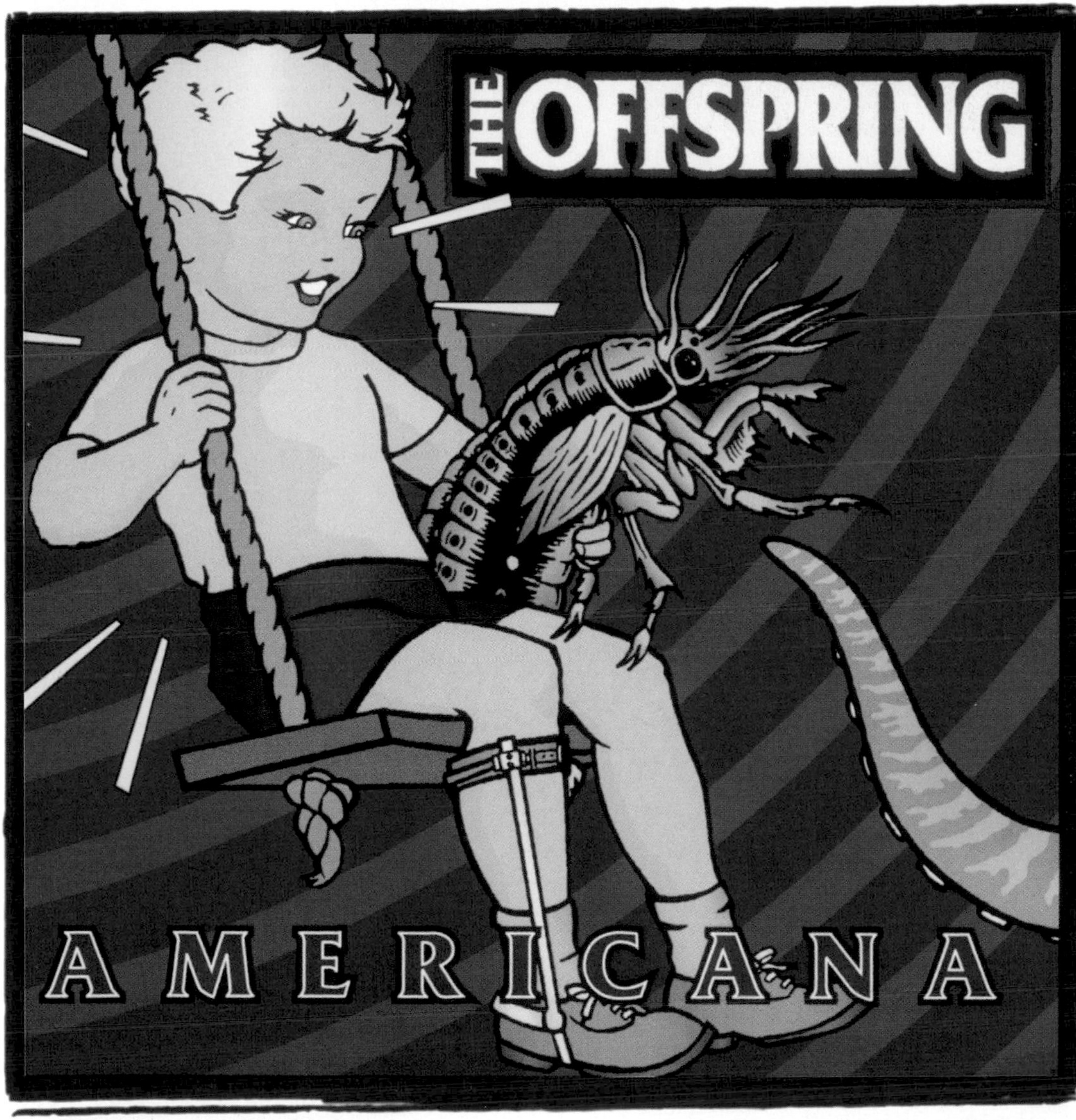
THE OFFSPRING
AMERICANA

BLINK-182

ENEMA OF THE STATE (1999)

Where the Offspring tried to capture some of the edginess of social commentary, Blink-182 went all-out for the puerile-humour jugular, which for a while enabled them to have their finger on the pulse; 1999 was also the year that saw the successful teen gross-out movie *American Pie* hit the screens. In keeping with such sex and orifice-based comedy, the **Enema Of The State** sleeve (referencing an 'enemy of the state', i.e. someone guilty of treason, and a phrase current at the time the album was recorded, thanks to the 1998 Will Smith film *Enemy Of The State*) features porn star Janine Lindermulder pulling a rubber glove on tight, as she presumably gets ready to perform the cleansing act. But then that was the whole hook for these types of gags: on the one hand you have the pleasure (*American Pie* saw the virgin male protagonist enjoying the feeling of having sex with an apple pie; here a porn star promises to be at a teenager's bidding), and on the other the embarrassment (the kid is caught by his father; the bidding is nothing more than a bowel-cleansing). Initial copies of the sleeve featured a Red Cross on Lindermulder's nurse's hat, until the Red Cross demanded its removal.

RECORD LABELS

MCA

RELEASE DATES

June 1999

SONGWRITERS

Tom DeLongue/Mark Hoppus

KORN

ISSUES (1999)

The downside of the success that bands such as Nirvana and Pearl Jam enjoyed – especially after Kurt Cobain's suicide turned him into an icon for the disaffected – is that self-loathing became big business in the late 90s. As metal bands such as Marilyn Manson and Cradle Of Filth faced increasing pressure from a media whipping up moral panic surrounding their influence (or not) upon record-buying teenagers, a degree of self-pity also became the norm, with many bands feeling the need to vent personal anguish from the position of the Great Misunderstood. There were actually four different sleeves for **Issues**, designed by fans for an MTV competition. The most famous, the slashed stuffed doll with a missing eye (an inspiration for a promotional doll later sold by the band), is symbolic of the overall self-loathing contained in the album's lyrics. Dashed childhood dreams, raging adolescence and disaffected youth are all symbolized in the image of the doll, seemingly assaulted like a murder victim. It is an image that uncomfortable teenagers bought into wholesale in the late 90s.

RECORD LABELS

Immortal

RELEASE DATES

November 1999

SONGWRITERS

Reginald 'Fieldy' Arvizu, Jonathan Davis, James 'Munky' Shaffer, David Silviera, Brian 'Head' Welch

ALBUM COVERS

NOUGHTIES

*NSYNC

NO STRINGS ATTACHED (2000)

In the wake of Take That's blistering success in the 90s, many boybands had a hard time being taken seriously – especially after a series of damaging 'revelations' concerning the groups' manufactured image and inability to write songs or even sing on their own records (forget performing live without miming). As far as boybands went, *NSync were actually at the forefront of the 'can stand on their own two feet' groups; they harboured a young Justin Timberlake, what more proof do you need? **No Strings Attached** reflects all this and more, as the group revelled in proving to everyone that they were able to do things on their own terms. The album also marked an upturn in *NSync's fortunes after their bitter split from RCA to newly sign with Jive, a label best known for its work with hip-hop acts, but also home to the most successful teen pop acts of the time, including Backstreet Boys and Britney Spears. It was something of a jibe at other boybands, whose images and output were seen to be more or less directed by faceless managerial puppeteers, rather than by the bands themselves.

RECORD LABELS

Jive

RELEASE DATES

March 2000

SONGWRITERS

Principal songwriters: Andreas Carlsson, J.C. Chasez

Secondary songwriters: Kevin Antunes/Reginald Calloway/Vincent Calloway/Teddy Pendergrass/Justin Timberlake, Kevin 'Shekspere' Briggs/Kandi, Bradley Daymond/Alexander Greggs/Lisa 'Left Eye' Lopes/Inga Willis, Kristian Lundin/Jacob Schulze, Max Martin/Rami, Richard Marx, David Nicoll/Veit Renn, Diane Warren, Robin Wiley

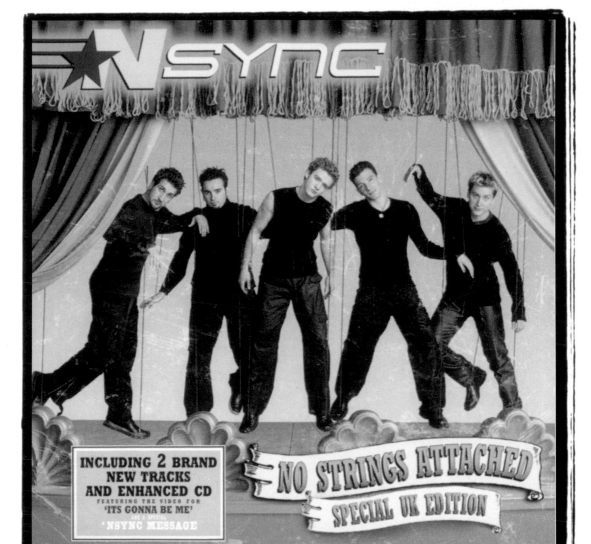

NO DOUBT

RETURN OF SATURN (2000)

American alt rock took a more poppy and glamorous turn in the 00s. No Doubt brought with them a post-post-punk revival stemming from the 90s, mixed with an eclecticism afforded alt-rock groups after hip hop and rock fully merged, thanks in no small part to Beck's **Odelay** (1996) and Beastie Boys' continued success. It wasn't until 1995's **Tragic Kingdom** that the band picked a sleeve devoid of the throwaway imagery of many other alt-rock designs. Though **Return Of Saturn** does not necessarily keep the classy feel of 1995's **Tragic Kingdom**, it retained the sense of kitsch and introduced the more freewheeling band that No Doubt had become, while also revealing the more mature character Gwen Stefani had become over the preceding five years. The album's title refers to the 'Saturn return' of a person's life – in astrological terms, the time when Saturn realigns itself with the position it occupied when a person was born, bringing with it the fears and anxieties that the planet is known for, while also marking maturity and reflection. Hence the mish-mash of bright, youthful images juxtaposed with the more mature, growing-up imagery of lipstick and wedding cake – and a large telescope, just right for watching the planet come round the corner.

RECORD LABELS

Interscope

RELEASE DATES

April 2000

SONGWRITERS

Tom Dumont, Tony Kanal, Gwen Stefani

348

NOUGHTIES

DIXIE CHICKS

HOME (2002)

For much of their career, Dixie Chicks fostered an innocuous, country-pop image, easily overlooked by anyone other than fans or country aficionados. Their 1992 debut, **Little Ol' Cowgirl**, pictured a young blond child with a wagon wheel bigger than herself, going some way to suggesting that the Chicks would happily indulge in self-effacement if it got them noticed. 1998's **Wide Open Spaces** presented a more mature group, earthier than Shania Twain, but making as much use of a pretty-girls-gone-country-pop look to get themselves into record buyers' homes. By 2002, however, country music had reached new levels of popularity, thanks to the Cohen Brothers' blockbuster *O Brother, Where Art Thou?* movie, Johnny Cash's image revival and Gillian Welch's emergence as a bluegrass artist for modern times. Dixie Chicks returned to country's bluegrass roots on **Home**, revealing their new, retrospective look in 40s/50s Grand Ole Opry dress on a sepia-toned sleeve. The album's title would prove ironic, given Natalie Maines' comment that the Chicks were 'ashamed the President of the United States is from Texas' while touring the UK in 2003. This resulted in mainstream radio and TV support for the group being almost entirely withdrawn in their homeland, just as they enjoyed some of their biggest exposure.

RECORD LABELS

Columbia, Monument

RELEASE DATES

August 2002

SONGWRITERS

Principal songwriters: Martie Maguire, Natalie Maines, Emily Robinson

Secondary songwriters: Radney Foster, Patty Griffin, Terri Hendrix, Gary Nicholson/Tim O'Brien, Stevie Nicks, Bruce Robinson, Darrell Scott, Maia Sharp/Randy Sharp

DIXIE CHICKS
★★★★★★★★★★★★★★★
Home

| 0 | 25 | 50 | 100 | 150 |

Copyright by DIXIE CHICKS & Co., T.X.

HOME

CHRISTINA AGUILERA

STRIPPED (2002)

The main pop rivalry in the early 00s was the battle between Britney Spears and Christina Aguilera for the Queen of Pop crown. Image-wise, Aguilera blew Spears out of the water with 2002's **Stripped**. Though Spears was making tentative steps towards sexing herself up a little, she ultimately kept it safe for the kids, never wholly chucking the girl-next-door look that appeased the parents who bought her records for their children. Aguilera's second major album, however, destroyed the nice girl image she fostered on her self-titled 1999 debut. The former *New Mickey Mouse Club* star (something Spears also boasted on her CV) revealed an image pushed to sexual overdrive: cowgirl jeans and her bare breasts covered only with long, tousled hair. It was something of a shock for the more conservative pop buyers, and with singles such as 'Dirrty' much of the focus fell on Aguilera's controversial, headline-grabbing image. It is ironic, then, that time has proven Aguilera the more capable of reinvention, having updated her appearance regularly, never relying on this overly sexed-up look for too long. Spears, however, has seen her image wholly tarnished, thanks to a series of increasingly bizarre public outbursts and drink and drug allegations. Funny how things change.

RECORD LABELS

RCA

RELEASE DATES

October 2002

SONGWRITERS

Principal songwriters: Christina Aguilera/Matthew Morris/Scott Storch

Secondary songwriters: Glen Ballard, Rob Hoffman/Heather Holley, Alicia Keys, Steve Morales/Balewa Muhammad/David Siegel, Linda Perry

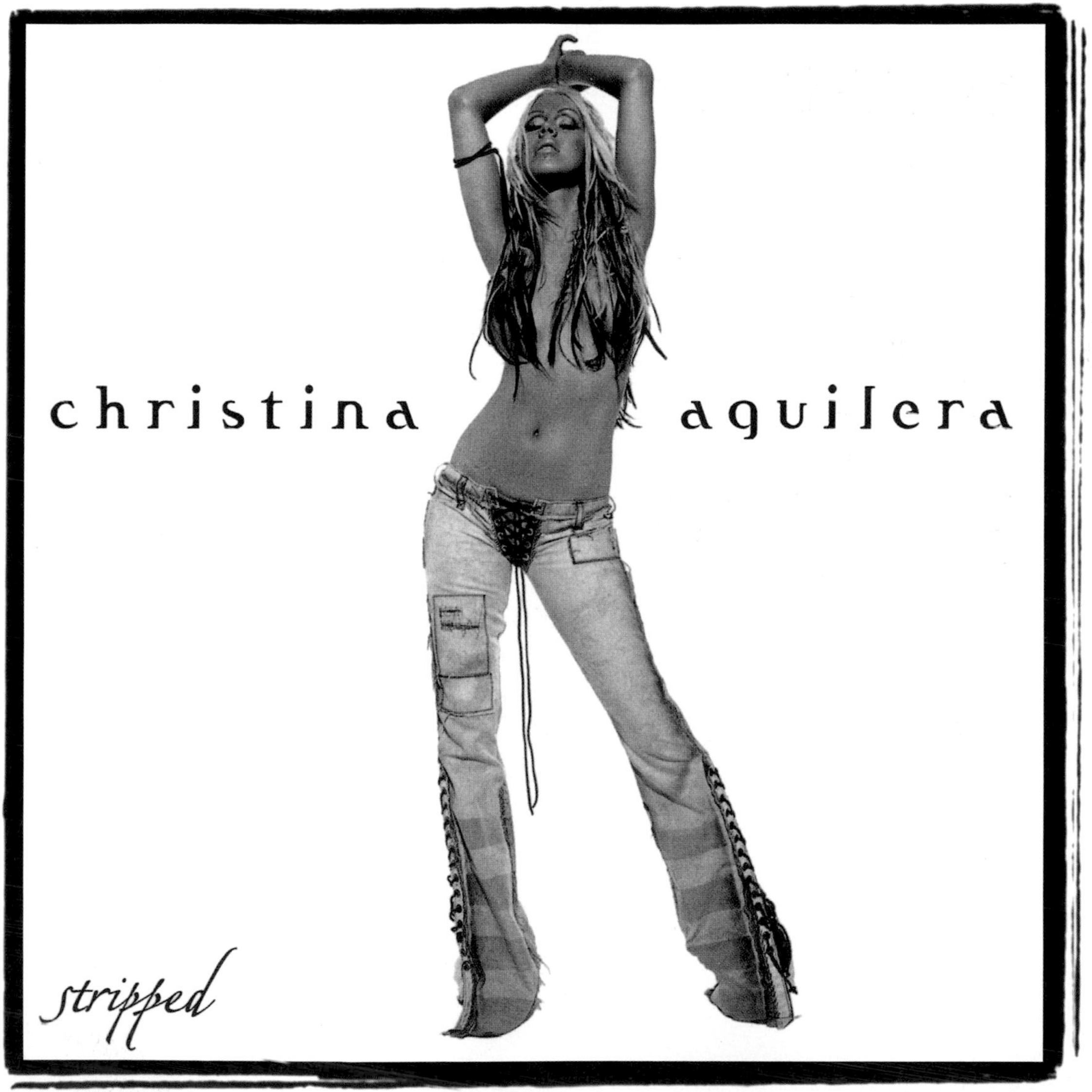
christina aguilera
stripped

SYSTEM OF A DOWN
STEAL THIS ALBUM! (2002)

Given the highly publicized battle between the recording industry and peer-to-peer file sharers that began to really take hold in the early 00s, it was only a matter of time before a band directly addressed it in their artwork. A collection of unreleased material, some of which had already been leaked onto the Internet following the release of the band's **Toxicity** (2001), **Steal This Album!** came in a jewel case with no booklet or proper artwork, just a CD and rear sleeve that both looked as if they had been scrawled on in magic marker. Essentially, it looks like just the sort of thing any kid would have downloaded and burned onto CD at the time, with the 'steal this' sentiment coming from a major label band five years before Nine Inch Nails' Trent Reznor was imploring his fans to illegally download his music. Extremely limited edition versions of the albums were pressed up for the United Kingdom and America, seeing four different CD designs made to look as though they had each been hand-drawn by a member of the band.

RECORD LABELS

Columbia, American

RELEASE DATES

November 2002

SONGWRITERS

John Dolmayan, Daron Malakian, Shavo Odadjian, Serj Tankian

THE WHITE STRIPES

ELEPHANT (2003)

The White Stripes' red-white-and-black self-branding is now so ingrained in the public consciousness that it is hard to remember a time when they weren't global superstars and simply released limited-run 7-inches in Detroit. **Elephant** made them household names, however, their critical and commercial appeal reaching its highest point with its release. Such popularity allowed them more freedom with their releases and they produced six different sleeves for **Elephant**, for CD and LP releases across the world. At this point in their career, the ever-symbolism-keen Jack White took the band's look up a notch, and the sleeve sees them dressed as Grand Ole Opry-style sweethearts. The title was chosen because of the way in which the Stripes perceived the animal: noble, mating for life and only attacking you if you threatened the young. White later went on to say of the sleeve: 'If you study the picture carefully, Meg and I are elephant ears in a head-on elephant. But it's a side view of an elephant, too, with the tusks leading off either side…. I wanted people to be staring at this album cover and then maybe two years later, having stared at it for the 500th time, to say, "Hey, it's an elephant!"'

RECORD LABELS

XL, V2

RELEASE DATES

April 2003

SONGWRITERS

Principal songwriter: Jack White

Secondary songwriters: Burt Bacharach/Hal David, Mort Crim

BLUR

THINK TANK (2003)

By the time Blur recorded their seventh album, Britpop, its hangover and the heartbreak album had all come and gone for Damon Albarn. He was enjoying the fruits of new project Gorillaz, which would bring him more global success than he had ever achieved with Blur. Unsurprisingly, tensions increased between Albarn and guitarist Graham Coxon, resulting in the latter leaving the band (he is present on only one song, the beautiful 'Battery In Your Leg') and the former running freely all over the record. If **Think Tank** can be seen as a somewhat paranoid piece of work at times, it is only fitting that Albarn enlisted Britain's most famous graffiti artist Banksy to create the art. Another example of how Gorillaz opened up Albarn's sole control over Blur, Banksy's infamous monkey image had previously appeared in the Gorillaz video for 'Tomorrow Comes Today', with the sign, 'Laugh now but one day we will be in charge'. Banksy's subversive artworks are sometimes cynical commentaries on modern living, in keeping with many of the album's themes, and are often preserved on the walls of buildings he paints them on. Demand for his paintings is so high that the **Think Tank** art sold for £75,000 in 2007.

RECORD LABELS

Parlophone

RELEASE DATES

May 2003

SONGWRITERS

Principal songwriters: Damon Albarn, Alex James, Dave Rowntree

Secondary songwriters: Graham Coxon, James Dring, Mike Smith

GORILLAZ

DEMON DAYS (2005)

If their first album was a toe in the water, Gorillaz' second dived headlong in. Masterminds Damon Albarn and Jamie Hewlitt took the ultimate manufactured pop band's sound and image to new heights, creating a peerless work rich in postmodernism (Gorillaz' story closely mirrors many of popular music's best-known tales), arrogance and the limitless possibilities of what can be done with a band that is truly created by others. The **Demon Days** sleeve is a take on the Beatles' **Let It Be**; as with the Beatles, who caricatured themselves and their differing personalities their entire career, **Demon Days** introduces the shifty Murdoc, rather sullen and gormless 2D, ultra-stylish Noodle and strong hip-hop backbeat Russell, using Manga-style illustration. A limited edition of the sleeve packaged the album in a four-panel fold-out digipak, which allowed the buyer to pick one larger image of an individual band member for their cover, and revealed an alternative image of each under the flaps.

RECORD LABELS

Parlophone

RELEASE DATES

May 2005

SONGWRITERS

Principal songwriters: 2D, Russell Hobbs, Murdoc Niccals, Noodle

Secondary songwriters: Daniel Dumile, Don Harper, David Jolicoeur, Romye Robinson, Rodney Smith/Simon Tong

GORILLAZ

PARENTAL
ADVISORY
EXPLICIT CONTENT

DEMON DAYS

HARD-FI

STARS OF CCTV (2005)

Taking their cue from the White Stripes' approach to bold colouring, Hard-Fi at least had the good sense to change the colour scheme when they decided a coded look was perfect for their image as well. Yellow and black signifies danger in the animal kingdom and, as far as Britain is concerned, it has similar connotations on road signs and barriers. Introducing the band as one to keep an eye on, then, the title says it all – these guys are delinquents, and proud of it (an alternate, hand-drawn sleeve made the gritty street connection even more pronounced). If there is one thing the 00s will be remembered for, it is that reality TV made stars of everyone (well, to some sort of degree at least), with contestants in the first-ever *Big Brother* gameshow of 2000 still making newspaper headlines in the late years of the decade. Hard-Fi are more or less saying they are infamous, then, playing to the cameras that presumably catch them peeing in shop doorways and emptying rubbish bins in the street. Certainly eye-catching, it worked better than the art for their second album, which boldly proclaimed 'NO COVER ART'. No, that would be the Beatles' **The Beatles** (see page 63).

RECORD LABELS

Warner Bros., Necessary/Atlantic

RELEASE DATES

July 2005 (US: March 2006)

SONGWRITERS

Richard Archer

362

FRANZ FERDINAND
YOU COULD HAVE IT SO MUCH BETTER (2005)

After the likes of the White Stripes and the Strokes kicked down the door to a new breed of stripped-down rock'n'roll groups, Franz Ferdinand headed up the British end with the umpteenth new wave of new wave. The Stripes have a lot to answer for, given the way in which their colour-scheme fixation took off, and Franz Ferdinand were almost carried along in the current for a while. When addressing their second album, mainman Alex Kapranos said, 'The whole point is that the album doesn't have a title. We decided quite a while ago that we didn't want to give any of the albums titles, they were just going to be called *Franz Ferdinand* [as their debut had been]…. The albums are going to be identified by their colour schemes rather than a title. The contrast of different colours creates a different mood. We experimented with different combinations of colours and this one stuck. At one level they looked good together, and they capture the mood of this record quite well.' However, they changed their mind at the last moment, titling it **You Could Have It So Much Better** and adding a Russian art-inspired design, giving their sleeve more depth than many of their contemporaries would have allowed.

RECORD LABELS

Domino

RELEASE DATES

October 2005

SONGWRITERS

Alex Kapranos, Bob Hardy, Nick McCarthy, Paul Thompson

FRANZ
FERDINAND

Domino

ROBBIE WILLIAMS

INTENSIVE CARE (2005)

Since leaving Take That in 1995, Robbie Williams' image has changed drastically from the carefully nurtured boyband look he embodied at that time. Almost immediately after leaving the group he started hanging around with the likes of Liam Gallagher, and appeared on *Shooting Stars* as a slightly overweight, possibly inebriated guest. It is fair to say that Williams is a man who has had his share of problems and also that, admirably, he doesn't shy away from them, seeming all too happy to face up to his demons. The sleeve for **Intensive Care**, the title itself a reference to hospitalization, plays out an almost intergalactic 'good versus evil' battle for Robbie's soul, with Angel Robbie (oddly dressed in charcoal, perhaps questioning whether there is such a thing) on one shoulder and Devil Robbie on the other. Williams himself seems to be reaching out, pressing down onto the buyer's assumed forehead in an almost new-age act of healing. The runic symbol alone suggests that Williams has come through a darker period to find a degree of humble spirituality. Compare this with the self-aggrandizement of **Swing While You're Winning** (2001), or the smug **I've Been Expecting You** (1998); it is like two different people.

RECORD LABELS

Chrysalis

RELEASE DATES

October 2005

SONGWRITERS

Principal songwriters: Stephen Duffy/Robbie Williams

Secondary songwriter: C.S. Heath

MADONNA

CONFESSIONS ON A DANCEFLOOR (2005)

The problem with the 80s stars, who came to fame in the Me Decade, is that they are only just dealing with growing old. By now the Bob Dylans and Neil Youngs have returned to their massive homes, turning out albums as they please, while David Bowie has fashioned himself into a guardedly affable character, popping out whenever he likes. Prince, however, deals with middle age by turning to Jehovah and refuses to let people photograph him, preferring instead to attack his fan websites for posting images of him, and Madonna tries to make it look as though she isn't growing old at all. For **Confessions On A Dancefloor** she gave herself an entirely neo-80s look for the (albeit partially indebted to the 80s) mid-00s, all flicked hair, leggings and neon colours. Was she reminding us that she did it all first? Mostly it just looks like the old dog can't be taught new tricks, something borne out by the album, actually more a return to kitsch 70s disco than 80s electro ('Hung Up' is only the second song Abba have given clearance for sampling). It might have reminded her fans of their youth, but it also reminded us of Madonna's age.

RECORD LABELS

Warner Bros.

RELEASE DATES

November 2005

SONGWRITERS

Principal songwriter: Madonna

Secondary songwriters: Mirwais Ahmadzaï, Benny Andersson/Björn Ulvaeus, Peer Astrom, Anders 'Bag' Bagge, Joe Henry, Henrik Jonback, Christian Karlsson, Pontus Winberg, Stuart Price

ARCTIC MONKEYS

WHATEVER PEOPLE SAY I AM, THAT'S WHAT I'M NOT (2006)

For a while the fastest-selling debut album in UK history, **Whatever People Say I Am, That's What I'm Not** ushered in Arctic Monkeys (particularly chief songwriter Alex Turner) as gritty social commentators for the 00s. It also introduced the world at large to the idea of word-of-mouth frenzy through the power of the Internet, even if such things have been slightly over-egged. The urban photorealism of Turner's lyrics is mirrored by the sleeve, which features Chris McClure, brother of fellow Sheffield musician Jon McClure, photographed in a black-and-white close-up, as taken during part of a long night out at Liverpool's Korova bar, after he and his cousin were given £70 to go out and enjoy themselves. In an era where the smoking ban had already reached Ireland, with England soon to follow suit, Scotland's NHS claimed that the sleeve glorified smoking. A spokesperson for Domino, the band's label, however, denied the claims, countering that, 'You can see from the image smoking is not doing him the world of good'. For American billboards a variation was used, with the cigarette removed. It is not inconceivable that this might be the last major-selling album to feature someone smoking on the cover.

RECORD LABELS

Domino

RELEASE DATES

January 2006 (US: February 2006)

SONGWRITERS

Principal songwriter: Alex Turner

Secondary songwriter: Jamie Cook

THOM YORKE

THE ERASER (2006)

Although **The Eraser** was Thom Yorke's first solo album away from Radiohead, he still enlisted longtime Radiohead sleeve designer Stanley Donwood to create the artwork, based on the *London Views* series of designs that Donwood began in 1995. The original art was carved out on 14 pieces of linoleum and folds out across five double-sided panels of cardboard sleeve packaging, depicting a series of famous London buildings, including the 'Gherkin', the Tower of London and the Houses of Parliament, being swept away by the River Thames. The Eraser figure stands by like King Canute the Great, the Viking king of England, Denmark, Norway and parts of Sweden, who attempted to hold back the sea's tide. In the promotional campaign running up to the album's release, a life-size model of the Eraser figure was placed at iconic London landmarks during selected times of the day, equipped with a pair of headphones that played the album on loop. The www.theeraser.net website invited fans to post photos of themselves posing with the Eraser if they found him, while the artwork was also painted across the XL label's building.

RECORD LABELS

XL

RELEASE DATES

July 2006

SONGWRITERS

Principal songwriter: Thom Yorke

Secondary songwriter: Jonny Greenwood

OASIS

STOP THE CLOCKS (2006)

Oasis' inventive sleeves took a bit of a downturn in the first half of the 00s until, ever the Beatles enthusiasts, they enlisted **Sgt. Pepper's Lonely Hearts Club Band** sleeve designer Peter Blake to come up with a hugely creative cover for their first 'best of'. Though the original sleeve was going to be a photo of famous 60s clothing shop Granny Takes A Trip (later the SEX boutique in the 70s, where the Sex Pistols would come together), Blake has said that he intended for the used design to be something of a meeting between **Sgt. Pepper...** and the sleeve for Oasis' debut, **Definitely Maybe** (see pages 49, 317). Blake claims to have picked all the items for **Stop The Clocks** at random, while, in a reference to **Sgt. Pepper...**, he included famous faces such as the Seven Dwarves, Dorothy from *The Wizard Of Oz* (1939) and Michael Caine (who replaced Marilyn Monroe after Blake was stopped from using her image). The overall effect is perfect for a compilation album, as it gives the feel of rummaging through an entire history of personal paraphernalia and nick-nacks. An actual **Stop The Clocks** dartboard was manufactured for promotion and is worth hundreds of pounds to Oasis collectors.

RECORD LABELS

Big Brother, Columbia

RELEASE DATES

November 2006

SONGWRITERS

Principal songwriter: Noel Gallagher

Secondary songwriter: Liam Gallagher

374

TOP CHART POSITIONS

The peak chart positions for every album featured are listed below in alphabetical order by artist, based on the official UK Album Chart and US *Billboard* Album Charts. A dash is used to indicate instances where a record did not hit the charts.

ARTIST	ALBUM	UK	US
AC/DC	Back In Black (1980)	1	4
Adam & the Ants	Kings Of The Wild		
	Frontier (1980)	1	44
Aerosmith	Get A Grip (1993)	2	1
Christina Aguilera	Stripped (2002)	2	2
The Allman Brothers Band	Eat A Peach (1972)	–	4
Arctic Monkeys	Whatever People Say I Am,		
	That's What I'm Not (2006)	1	24
Beastie Boys	Licensed To Ill (1986)	7	1
The Beatles	Rubber Soul (1965)	1	1
	Sgt. Pepper's Lonely Hearts		
	Club Band (1967)	1	1
	The Beatles (1968)	1	1
	Abbey Road (1969)	1	1
Beck	Odelay (1996)	18	16
Big Brother & the			
Holding Company	Cheap Thrills (1968)	–	1
Bjork	Debut (1993)	3	61
Black Crowes	Amorica (1994)	8	11
Black Grape	It's Great When You're		
	Straight … Yeah (1995)	1	–
Blink-182	Enema Of The State (1999)	15	9
Blondie	Parallel Lines (1978)	1	6
Blur	Think Tank (2003)	1	56
Bow Wow Wow	See Jungle! See Jungle!		
	Go Join Your Gang,		
	Yeah. City All Over!		
	Go Ape Crazy! (1981)	26	–
David Bowie	Aladdin Sane (1973)	1	17
	Diamond Dogs (1974)	1	–
	"Heroes" (1977)	3	35
Captain Beefheart			
& the Magic Band	Doc At The Radar Station (1980)	–	–
The Cars	Candy-O (1979)	30	3
Johnny Cash	American Recordings (1994)	–	110
The Clash	London Calling (1979)	9	27
Ornette Coleman	Change Of The Century (1959)	–	–
Alice Cooper	School's Out (1972)	4	2
Elvis Costello & the Attractions	Trust (1981)	9	28
Cream	Disraeli Gears (1967)	5	4
The Cure	Three Imaginary Boys (1979)	44	–
Miles Davis	Bitches Brew (1970)	71	35
	Tutu (1986)	74	–
Dead Kennedys	Fresh Fruit For Rotting		
	Vegetables (1980)	33	–
Depeche Mode	A Broken Frame (1982)	8	–
Dixie Chicks	Home (2002)	58	1
The Doors	Strange Days (1967)	–	3
Duran Duran	Rio (1982)	2	6
Ian Dury	New Boots And		
	Panties!! (1977)	5	–
Ian Dury & the Blockheads	Do It Yourself (1979)	2	–
Bob Dylan	Bringing It All Back Home (1965)	1	6
	Self Portrait (1970)	1	4
The Eagles	Hotel California (1976)	2	1
Emerson, Lake & Palmer	Brain Salad Surgery (1973)	2	11
The Faces	Ooh La La (1973)	1	21
Ella Fitzgerald	Ella And Louis (1956)	–	12
Fleetwood Mac	Rumours (1977)	1	1
Franz Ferdinand	You Could Have It So Much		
	Better (2005)	1	8
Funkadelic	Cosmic Slop (1973)	–	–
	Hardcore Jollies (1976)	–	96
Peter Gabriel	Peter Gabriel (1980)	1	22
Gorillaz	Demon Days (2005)	1	6
Grateful Dead	Aoxomoxoa (1969)	–	73
	American Beauty (1970)	–	30
Green Day	Dookie (1994)	13	2
Hard-Fi	Stars Of CCTV (2005)	1	–
Debbie Harry	Koo Koo (1981)	6	23
Isaac Hayes	Hot Buttered Soul (1969)	–	8
Iron Maiden	The Number Of The		
	Beast (1982)	1	33
It's A Beautiful Day	It's A Beautiful Day (1969)	58	47
Joe Jackson	Look Sharp! (1979)	40	20
Elton John	Goodbye Yellow Brick		
	Road (1973)	1	1
Grace Jones	Slave To The Rhythm (1985)	12	73
Quincy Jones	This Is How I Feel About		
	Jazz (1956)	–	–
Joy Division	Unknown Pleasures (1979)	–	–
King Crimson	In The Court Of The		
	Crimson King (1969)	5	28
Korn	Issues (1999)	37	1
Kraftwerk	The Man-Machine (1978)	9	–
Cyndi Lauper	She's So Unusual (1983)	16	4
Led Zeppelin	Led Zeppelin (1969)	6	10
	Houses Of The Holy (1973)	1	1
	Physical Graffiti (1975)	1	1
Little Feat	Sailin' Shoes (1972)	–	–
LL Cool J	Mama Said Knock You		
	Out (1990)	49	16
Lynyrd Skynyrd	Street Survivors (1977)	13	5
Madness	One Step Beyond… (1979)	2	–

Artist	Album		
Madonna	True Blue (1986)	1	1
	Confessions On A Dancefloor (2005)	1	1
Bob Marley & the Wailers	Catch A Fire (1973)	–	–
Massive Attack	Mezzanine (1998)	1	60
Dave Matthews Band	Before These Crowded Streets (1998)	–	1
George Michael	Faith (1987)	1	1
Midnight Oil	Diesel And Dust (1987)	19	21
Steve Miller Band	The Joker (1973)	–	2
Joni Mitchell	Hejira (1976)	11	13
The Mothers of Invention	Weasels Ripped My Flesh (1970)	28	–
Gerry Mulligan & Thelonious Monk	Mulligan Meets Monk (1957)	–	–
New Order	Power, Corruption And Lies (1983)	4	–
New York Dolls	New York Dolls (1973)	–	–
Nirvana	Nevermind (1991)	7	1
No Doubt	Return Of Saturn (2000)	31	2
*NSYNC	No Strings Attached (2000)	14	1
N.W.A	"Straight Outta Compton" (1989)	41	37
Oasis	Definitely Maybe (1994)	1	58
	Stop The Clocks (2006)	2	89
The Offspring	Americana (1998)	10	2
Ol' Dirty Bastard	Return To The 36 Chambers (1995)	–	7
Graham Parker & the Rumour	The Parkerilla (1978)	14	–
Pearl Jam	No Code (1996)	3	1
Pink Floyd	Atom Heart Mother (1970)	1	55
	The Dark Side Of The Moon (1973)	2	1
	Wish You Were Here (1975)	1	1
	Animals (1977)	2	3
Plastic Ono Band	Live Peace In Toronto 1969 (1969)	–	10
The Police	Ghost In The Machine (1981)	1	2
Iggy Pop	The Idiot (1977)	30	72
Elvis Presley	Elvis Presley (1956)	–	1
	50,000,000 Elvis Fans Can't Be Wrong (1959)	–	31
Primal Scream	Screamadelica (1991)	8	–
Prince	Dirty Mind (1980)	–	45
	Sign 'O' The Times (1987)	4	6
Prodigy	The Fat Of The Land (1997)	1	1
Public Enemy	Fear Of A Black Planet (1990)	4	10
Public Image Ltd	Metal Box (1979)	1	–
Radiohead	OK Computer (1997)	1	21
Rage Against The Machine	Rage Against The Machine (1992)	17	45
Ramones	Ramones (1976)	–	–
Red Hot Chili Peppers	Blood Sugar Sex Magik (1991)	25	3
R.E.M.	Document (1987)	28	10
The Rolling Stones	Sticky Fingers (1971)	1	1
	Exile On Main Street (1972)	1	1
	Some Girls (1978)	2	1
Roxy Music	Siren (1975)	4	50
Santana	Abraxas (1970)	7	1
Boz Scaggs	Middle Man (1980)	52	8
The Sex Pistols	Never Mind The Bollocks ... Here's The Sex Pistols (1977)	1	106
Carly Simon	Playing Possum (1975)	–	10
Frank Sinatra	Come Fly With Me (1958)	–	1
Sly & the Family Stone	There's A Riot Goin' On (1971)	31	1
The Small Faces	Ogdens' Nut Gone Flake (1968)	1	–
Patti Smith	Horses (1975)	–	47
The Smiths	The Smiths (1984)	2	–
	Meat Is Murder (1985)	1	–
Snoop Doggy Dogg	Doggystyle (1993)	38	1
Soft Machine	The Soft Machine (1968)	1	–
Sonic Youth	Daydream Nation (1988)	99	–
Spiritualized	Lazer Guided Melodies (1992)	27	–
Bruce Springsteen	Born To Run (1975)	17	3
	Born In The USA (1984)	1	1
Steely Dan	Aja (1977)	5	3
Rod Stewart	Never A Dull Moment (1972)	1	2
Supertramp	Breakfast In America (1979)	3	1
System Of A Down	Steal This Album! (2002)	56	15
Talking Heads	Fear Of Music (1979)	33	21
	Remain In Light (1980)	21	19
	The Name Of This Band Is Talking Heads (1982)	22	31
The Cecil Taylor Quartet	Looking Ahead! (1958)	–	–
The 13th Floor Elevators	The Psychedelic Sounds Of The 13th Floor Elevators (1966)	–	–
Toto	Turn Back (1981)	–	41
Traffic	The Low Spark of High Heeled Boys (1971)	–	7
U2	War (1983)	1	12
	The Joshua Tree (1987)	1	1
	Rattle And Hum (1988)	1	1
	Achtung Baby (1991)	2	1
Underworld	dubnobasswithmyheadman (1993)	12	–
Van Halen	1984/MCMLXXXIV (1984)	15	2
The Velvet Underground	The Velvet Underground And Nico (1966)	–	–
Tom Waits	Small Change (1976)	–	89
	Franks Wild Years (1987)	20	–
Scott Walker	Scott (1967)	3	–
The White Stripes	Elephant (2003)	1	6
The Who	The Who Sell Out (1968)	13	48
	Tommy (1969)	2	4
	Who's Next (1971)	1	4
Robbie Williams	Intensive Care (2005)	1	–
Stevie Wonder	Stevie Wonder's Journey Through The Secret Life Of Plants (1979)	8	4
Phil Woods & Donald Byrd	The Young Bloods (1956)	–	–
X	Los Angeles (1980)	–	3
Yes	Close To The Edge (1972)	4	3
Thom Yorke	The Eraser (2006)	3	2
Frank Zappa	Hot Rats (1969)	9	–
ZZ Top	Eliminator (1983)	3	9

ACKNOWLEDGEMENTS

AUTHOR BIOGRAPHIES

JASON DRAPER (AUTHOR)

Jason Draper is the Reviews Editor of *Record Collector* magazine and this is his second book for Flame Tree. A true collector himself he can attest to the visual allure of LP album sleeves, and the fact that allowing yourself to get sucked in by their charms will lead you down the rocky road of never being happy until you've got 'just one more'. Like all good addictions, the first one was free, but it probably got too far when records owned became greater than floorspace free.

Special Acknowledgement

Jason would like to thank his brother Matthew, who's changed his look quite a bit recently.

PAUL DU NOYER (FOREWORD)

Paul Du Noyer began his career on the *New Musical Express*, went on to edit *Q* and to found *Mojo*. He also helped to launch *Heat* and several music websites. As well as editing several rock reference books, he is the author of *We All Shine On*, about the solo music of John Lennon, and *Wondrous Place*, a history of the Liverpool music scene. Nowadays he is a contributing editor of *The Word*.

PICTURE CREDITS

The individual record companies, artists, photographers and designers retain the copyright for cover images. Special thanks to all record labels, management companies and artists, especially the following:

Apple Corps Ltd (© Apple Corps Ltd): 43, 49, 63, 77
Cherry Red Records: 219
Stanley Donwood (artwork by and © 2006): 373
EMI Music: 177, 205, 213, 217, 239, 243, 331, 337, 359, 361, 367
Leadclass Ltd: 137
Pink Floyd (1987) Ltd (All Pink Floyd album covers supplied courtesy of Pink Floyd [1987] Ltd): sleeve design by Hipgnosis: 95, 145; sleeve design by Storm Thorgerson and Hipgnosis: 119; sleeve design by Roger Waters organized by Storm Thorgerson and Aubrey Powell: 163
Promotone B.V.: 99, 111, 183
RZO Music: 127, 139, 173
Sony BMG Music Entertainment (UK) Ltd: 207, 225, 235, 281, 303
Stirling Holdings Ltd (licensed c/o, photo by Cameron McVey): 201
Studio One, New York/Mrs Yoko Ono Lennon: 83
The White Stripes/Third Man Records: 357

FURTHER READING

Aldis, N. and Sherry, J., *Heavy Metal Thunder: Album Covers that Rocked the World*, Mitchell Beazley, 2006

De Ville, N., *Album: Classic Sleeve Design*, Mitchell Beazley, 2005

De Ville, N., *Album: Style and Image in Sleeve Design*, Mitchell Beazley, 2003

Drate, S., *45 RPM: A Visual History of the Seven-Inch Record*, Princeton Architectural Press, 2002

Drate, S., *Rock Art: CDs, Albums and Posters*, PBC International, 1994

Kohler, E., *In the Groove: Vintage Record Graphics 1940–1960*, Chronicle Books, 1999

Marsh, G. and Lewis, B. (ed.s), *The Blues: Album Cover Art*, Collins and Brown, 1995

Morrow, C., *Stir It Up! Reggae Album Cover Art*, Thames and Hudson, 1999

O'Brien, T., *Naked Vinyl: Classic Album Cover Art Unveiled*, Universe, 2003

Ochs, M., *Classic Rock Covers*, Taschen, 2001

Ochs, M., *1000 Record Covers*, Taschen, 2008

Rivers, C., *CD Art: Innovation in CD Packaging Design*, RotoVision, 2003

Scott, M. and S., *The Greatest Album Covers of All Time*, Collins and Brown, 2005

Shaughnessy, A., *Cover Art By:*, Laurence King, 2008

Takumi, M., *In Search of the Lost Record: British Album Cover Art from the 50s to the 80s*, Gingko Press, 2002

Thorgerson, S., *Classic Album Covers of the 60s*, Collins and Brown, 2005

Thorgerson, S. and Powell, A., *100 Best Album Covers*, Dorling Kindersley, 1999

Walsh, G., *Punk on 45: Revolutions on Vinyl 1976–79*, Plexus Publishing, 2005

Wax Poetics, *Cover Story: Album Cover Art*, Powerhouse Books, 2001

Weidemann, J. (ed.), *Jazz Covers*, Taschen, 2008

INDEX BY ALBUM

GENERAL INDEX

Page references in **bold** indicate major articles; those in *italics* indicate illustrations. Hyphenated page references take no account of intervening illustrations.